PROJECT CONTROLS

F O R

FEDERAL

CONSTRUCTION

PROJECT CONTROLS

FOR

FEDERAL

CONSTRUCTION

A Strategic Planning Guide for Federal Construction Projects

Scott Arias, PhD & Matt Ellis, MS

**PERFORMANCE
PUBLISHING**

Performance Publishing Group
McKinney, TX

ISBN:
978-1-967451-05-0 (paperback)

CONTENTS

INTRODUCTION

Project management might be one of the most widely used job ti-
tles across all industries. Computer programmers have project managers. Advertising campaigns have project managers. Everything
down to your local non-profit organization utilizes project managers in one way or another. This book is not meant to minimize the
importance of what any of those roles contribute to their respective
organizations, but rather to detail and explain the complex role and
responsibilities of project managers in the construction industry –
specifically the world of projects for the federal government.

The role of a Project Manager (PM) involves the forecasting of time,
material, tools, and equipment. The Project Manager must also
keep a project on schedule and maintain profitability in the face of
multiple problematic situations that will undoubtedly arise no matter how well planned the project might be. The enormous workload combined with limited labor and materials complicate almost
every project. Faced with all of these challenges, there is also the
responsibility for safety and quality control. Most importantly, the
Project Manager and the Superintendent are leaders who must set
the example for their crew while maintaining a positive attitude.
Being a Project Manager is a difficult job, but our goal is to show
you it is not impossible.

The Project Manager interfaces daily with tradesmen, business executives, and both external and internal Customers. They serve as the primary point of contact for vendors and Clients alike. A successful Project Manager must be thoroughly familiar with supply chain logistics, scheduling, budgets, safety, quality, and construction processes. Project managers are responsible for disseminating and fully implementing the principles outlined in this project management guide and for adapting them as needed in the field.

This book is not meant to be a replacement or alternative for any professional construction industry certifications or project management training courses, but instead to work hand in hand with the technical knowledge you already have and serve as a manual for navigating the rigorous requirements that come with federal contracts. With that standard in mind, we also want this to be a benefit to anyone in the construction industry, no matter the size or scope of the projects you work on. From a broad level overview of the entire process and how to successfully bid on jobs, all the way down to efficiently closing out projects and finding a wealth of forms you can employ in your daily operations – this book has something for every construction professional out there written by experts in the field.

ACE Consulting was founded in 2008 by Dr. Scott Arias, a retired Navy Seabee who was one of the youngest to achieve the rank of Chief Petty Officer at the time of his promotion, and a former tenured professor in the Construction Management program at Eastern Kentucky University. We operate on the three core values of always doing the right thing, working with extreme urgency, and always striving to be the gold standard in the construction industry.

We only succeed when our Clients do, and we are bringing that same philosophy to you, our readers.

Scott still serves as the CEO of ACE Consulting today, bringing with him a wealth of knowledge and experience. He holds a PhD in Construction Management, maintains multiple industry-specific certifications and licenses, and has over three decades of industry experience. In addition to overseeing the strategic vision and growth for ACE, Dr. Arias has written and published two books outside of this one and teaches on an adjunct basis at Eastern Kentucky University. He retired as a Senior Chief Petty Officer after twelve years of service and continued his career by going on to oversee the construction of twelve US embassies overseas before ultimately settling in Nicholasville, Kentucky, to embark on the path leading to where he is today.

His co-author, Charles Matthew Ellis, has over seventeen years of experience as a project manager – fourteen of those at ACE, where he currently serves as the Company President. His well-rounded experience in residential, industrial, and commercial construction projects provide a unique perspective to anything a Project Manager might encounter on a government project. He also holds a Master's Degree in Construction Management from Eastern Kentucky University as well as other certifications and accreditations. His expertise in various industry specific software programs coupled with knowledge of both the project administration side of the business and environmental compliance make him the perfect complement to everything Dr. Arias and ACE Consulting bring to the table.

CHAPTER 1

CONSTRUCTION MANAGEMENT OVERVIEW

Construction management is the disciplined process of planning, coordinating, and controlling a project from inception to completion. It ensures that safety, quality, budget, and schedule objectives are achieved while balancing the needs of the Owner, design team, contractors, and other stakeholders. Effective construction management provides structure and training for all personnel involved — Project Managers (PM), Quality Control Managers (QCM), Project Executives (PE), and others — and equips them with the tools to keep projects on track. This foundation introduces Project Managers to the fundamentals of project management, which can be divided into the following phases:

1) **Preconstruction/Planning:** This phase establishes the foundation for the entire project. It begins with a thorough review of the plans, specifications, RFP, and project estimate. From this, the work is broken down into distinct activities, each representing a manageable portion of the overall

scope. Estimating occurs here as well, with all materials, tools, equipment (including safety gear), and manpower requirements identified and listed on a Construction Activity Summary Sheet. This careful planning ensures the project team enters the execution phase with accurate data, defined scopes, and realistic expectations for labor, cost, and schedule.

2) **Scheduling & Project Controls:** Once activities and resources are defined, they must be logically arranged. Tasks are sequenced from beginning to end to illustrate their interdependencies, forming a logic network. Based on this network, the construction schedule is developed, establishing projected start and finish dates for each activity. A critical path is identified to highlight those tasks that directly impact project completion if delayed. This schedule serves as the management roadmap, ensuring that focus remains on the activities most vital to timely completion while giving all stakeholders a clear view of project milestones.

3) **Execution:** Execution is where planning becomes action. The Project Manager and field leadership ensure that all necessary resources are mobilized and available at the site when required. Crews, subcontractors, and vendors are coordinated to perform the work in alignment with the schedule and budget. The Project Manager is responsible for supervising daily operations, promoting safety, and maintaining quality standards. This phase is where leadership, communication, and problem-solving skills are most critical, as field conditions often require adaptation without compromising overall project objectives.

4) **Monitoring & Control:** No plan succeeds without active oversight. Monitoring and control involves continuously tracking progress, verifying that work is performed according to the schedule, and making necessary adjustments. The three-week lookahead schedule becomes a key tool, breaking down the master schedule into a short-term, detailed view of upcoming activities. Completed items fall off, while new ones are added each week. Resource allocation — labor, equipment, and materials — is actively managed to prevent shortages or delays. This ongoing process ensures the project stays on schedule, within budget, and aligned with safety and quality goals.

5) **Closeout:** The closeout phase transitions the project from active construction to Owner turnover. Work is finalized, punch lists are completed, and systems are tested and commissioned to confirm performance. Documentation is gathered, including as-builts, warranties, and close-out reports, to provide the Owner with a complete record of the project. Final inspections are scheduled with the Client to secure acceptance. Effective closeout not only ensures compliance with contract requirements but also strengthens Client relationships by demonstrating professionalism and accountability through the project's completion.

OPERATIONS STRUCTURE

An understanding of the operations structure and its responsibilities in the planning and execution of tasking is necessary prior to any further discussion of project management. The organizational chart below offers a graphical representation for what a typical operations structure within a construction project looks like.

Operations Organization

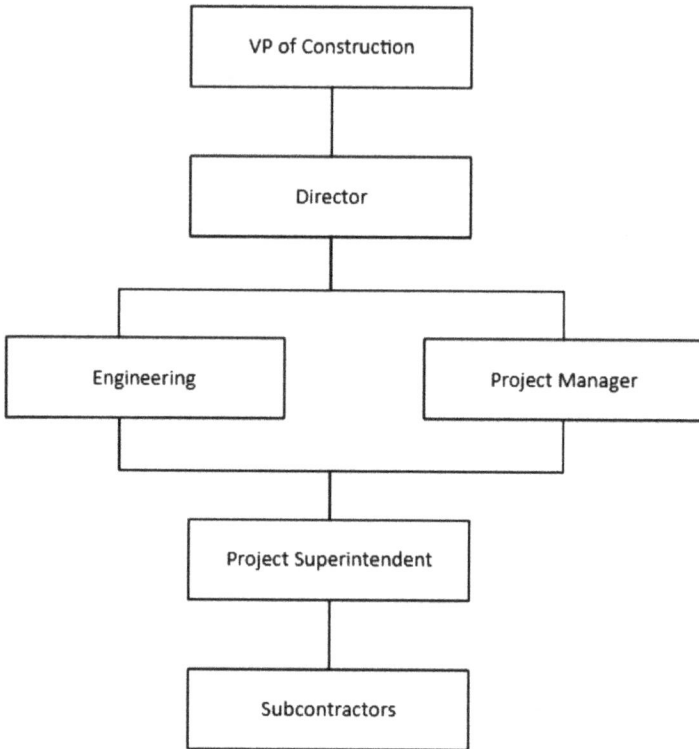

```
            ┌─────────────────────┐
            │  VP of Construction  │
            └─────────────────────┘
                       │
            ┌─────────────────────┐
            │      Director        │
            └─────────────────────┘
                       │
        ┌──────────────┴──────────────┐
┌───────────────┐            ┌──────────────────┐
│  Engineering  │            │  Project Manager  │
└───────────────┘            └──────────────────┘
        │
┌───────────────────────┐
│ Project Superintendent │
└───────────────────────┘
            │
┌───────────────────────┐
│     Subcontractors     │
└───────────────────────┘
```

At the top of the hierarchy is the VP of Construction. The VP has functional and administrative authority over all construction processes, some of which include estimating, project management, and safety. This person has the responsibility for the strategy and assigned resources to accomplish the group's corporate mission.

The Director reports directly to the VP of Construction and has functional authority over all construction processes, some of which include project management and safety. This person has the responsibility for assigned resources to accomplish the group's mission and directing the daily operations of construction, while directly overseeing the Engineering department and the Project Manager.

The Project Manager has administrative and functional authority over every project he is assigned. In addition, he is the record keeper for the project and is responsible for all administrative functions for the people he supervises. In addition to overseeing administrative functions, the Project Manager shares responsibility for adhering to safety, quality, and timeliness guidelines for the project along with the Superintendent.

The Architectural/Engineering (A/E) Design Manager has functional authority over the architecture and engineering of a project. These responsibilities include drafting, revising, and providing technical expertise to the field. This individual is responsible for specifying every single material component of a project, from the make/model of every mechanical system right down to the individual nuts and bolts used to secure everything in place.

The Project Superintendent/Site Manager has functional authority over all construction processes on their project, upon commencement of work on the jobsite. In addition this superintendent is responsible for all administrative actions involving the people they supervise. They share in responsibility for adhering to safety, quality, and timeliness guidelines for the project along with the Project Manager.

PRELIMINARY PLANNING

Preliminary planning for construction begins immediately upon receipt of an award from the Client. Detailed project planning is conducted during the course of submittal and review, but prior to initiation of any scope of work. There are many facets of preliminary planning to take into account in order to make sure the project is set up for success. Here are some of the key considerations every preliminary planning phase should contain:

- Begin working on letters of intent, subcontracts, purchase orders, supplier agreements, etc.
- Identify long lead items and procure submittals for those ASAP.
- Begin reviewing resource allocations.
- Review cash flow to determine how the project will be initiated, as the first pay application may not be due for some time.
- Start the process of identifying the current utility locations by contacting the local municipal authority or hiring a private Company.
- Start process of connecting temporary utilities and constructing any temporary facilities needed.
  - water
  - electricity
  - portable restrooms
  - jobsite trailers
  - laydown areas for storage of material and equipment
- Complete subcontractor buyout or finalize procurement.

DETAILED PROJECT PLANNING

The Project Manager compiles a five-folder project package. A sequence of planning steps is provided in this chapter. Through the project planning stages each section of the project package is created. An example of a sample project package can be found in Appendix 1-1, but we will cover a high-level overview of the key items to be aware of here. Each of these items will be covered in greater detail throughout the book as we approach the appropriate phase of construction as well, so should you want to jump ahead to specific sections, use the Appendix and this summary as your guide.

The first folder contains items related to general project information, correspondence, and financial documents. This is where you will find the original signed contract and scope sheet, ensuring there is no confusion about authority and limitations. The RFP, Estimating Worksheet, and Schedule of Values are also included to provide a reminder of how the project was awarded and structured. All notes and records of correspondence regarding general matters and scheduling belong here, along with invoices associated with the material and labor required to start and maintain the project.

The second folder contains all documents related to project activities. This includes Situation Reports, schedules, summary sheets, and any type of progress or activity reporting. These documents provide the project team with a clear record of day-to-day and milestone activities throughout the project.

The third folder focuses on project resources. It includes documentation on tools, equipment, and materials, along with any special notes regarding lead times, delivery schedules, or circumstances that may affect the availability of these items. This folder serves as the reference point for resource management during the project.

The fourth folder is dedicated to safety and changes. All safety requirements and specifications for protecting the job site are outlined here, including documents related to environmental concerns such as hazardous materials and waste. The "change" portion of this folder captures the entire change process, from requests for clarification and justification for changes to the associated approvals. Keeping these records organized ensures that safety compliance and contract modifications can be quickly referenced during construction.

The fifth folder contains the project's technical documents: plans, drawings, and specifications. These materials serve as the foundation for all work performed in the field. The documentation in this folder is closely linked to the change records in Folder #4, ensuring that any modifications to the project are reflected in the most current version of the plans and specifications.

ASSIGNING PLANNING AND ESTIMATING RESPONSIBILITIES

The Design Manager working with the Project Manager must assign project planning responsibilities. The project manager will develop a milestone worksheet from input he has received from the design and construction team. This milestone worksheet will assign responsibilities and completion date requirements to each member of the project team. This could mean assigning Project Managers to individual tasks or specialties, such as by trade, so that no one person is overwhelmed or runs the risk of missing a key date. Deliverables placed on a timeline are key to project bidding success.

REVIEWING PLANS AND SPECIFICATIONS

The next step in project planning is a thorough review of the plans and specifications. Below is a brief checklist to assist with this review along with key points to keep in mind:

- Scheduling
  - What is the scheduled progress?
  - Is sufficient time allotted?
  - Will long lead materials be available?
  - Is work required in an occupied building?
  - What are the local weather conditions?
  - Is phasing of the work necessary?
- Site Conditions
  - Are there hazardous materials?
  - Is debris removal specified?
  - Is there a staging area?
  - What other contractors will be working in the area?
  - What clearances are required to enter the jobsite?
  - What permits are required for the jobsite?
- Methods
  - Could any specific methods be more costly than expected?
  - Does the crew have necessary skills to perform the work?
  - What tools will be required and what can be rented on-site?

Identifying items with long lead times is critical early in the project. Practical timelines must account for expected material arrival dates. There is no fixed duration that automatically defines a "long lead" item — the key factor is whether a delay in delivery would affect an activity on the critical path. Project teams should never assume an item will be readily available simply because it has been in the past, especially when tied to critical activities.

Any special training requirements must be coordinated with human resources and confirmed early enough to avoid delays. Following up to ensure training is scheduled and completed in advance is essential to maintaining project momentum and meeting schedule commitments.

CHAPTER 2

BIDDING AND ESTIMATING

The responsibility for the bidding and estimating can vary based on Company size. Smaller companies will often have one Project Manager who handles all aspects from bidding to project closeout. Larger companies have dedicated estimators who are responsible solely for doing quantity take offs, gathering subcontractor numbers, adding applicable overheads, taking tools and equipment into account, and reconciling costs before submitting the bid. If the job is awarded, these companies will have a turnover meeting or turnover process to a Project Manager.

Regardless of who handles this aspect, they are ultimately responsible for ensuring the profitability of the project. All required resources are listed on the Construction Activity Summary Sheet (CASS), which will be reviewed in more detail at the end of this chapter. The scheduled start and finish dates are taken from the CASS. The assets are then connected to the timeline and any steps needed to monitor or obtain resources can be tracked on the CASS.

Estimating is not just getting coverage during the bidding stage, but about ensuring the scopes of the subcontractors and vendors are clear and encompass the entire scope of work. The Project Manager is responsible for driving the estimating process and keeping people accountable for the bids and subsequent negotiation.

ACTIVITY LISTINGS

Before diving deeper into our project planning, we'll split our project into smaller chunks and estimate each part separately. We'll start by dividing the project into eight to ten main activities (Summary Activities), each representing a large, functional section. Then, we'll break down each main activity into one to fifty more specific construction tasks. This approach forms the foundation for creating a well-defined scope. This is the basis for creating a well-thought-out scope.

Master Activities

The Project Manager assigns master activities to the projects. The master activities are based on CSI divisions (Concrete, Earthwork, Electrical, HVAC, etc.) and broken down into about five to ten construction activities that represent functional parts of the facility and are typically associated with specific trades. To support effective planning, it's essential that all personnel clearly understand the scope of each master activity; this is the primary purpose of the master activity listing. Providing a clear narrative description for each master activity ensures that everyone knows exactly where each work element belongs. This will reduce the chance of omitting any items of work when estimating.

Construction Activities

The Project Manager must break each master activity down into construction activities. A typical project might include fifteen to fifty such activities. For tracking purposes, construction activities are assigned simple numeric codes. These codes typically use the first digit(s) to represent the master activity (aligned with the CSI Division) and the last digit(s) to designate the individual activity. For example, an activity coded as 210 would indicate demolition (CSI Division 2) with "10" signifying the specific task, such as the demolition of a structure.

MATERIAL ESTIMATING

Before deciding on the resources needed for each building task, we need to pick our construction approach. Our choice of method and how we direct our contractor to offer the most competitive bid will impact the quantity of materials, equipment and tool types, safety gear, and workforce needs. The method chosen should be identified on the CASS in the assumptions block. It is imperative this occurs so the contractor gets guidance and the scopes of work closely mirror each other, making the selection process during buy-out even easier.

The project manager must do a material take off for each activity. One method of doing this is by highlighting each item on plans to ensure no item is omitted or counted twice. The project manager must also compare the tool and equipment requirements for each activity to ensure the subcontracts use the same method for each scope of work. Method continuity allows you to accurately compare scopes from a multitude of contractors. The cost of any unique tools or equipment rental will have to be included to ensure we are directing the subcontractors in the most economical direction they can go.

MAN-DAY ESTIMATES AND DURATIONS

A man-day is simply the unit of time it takes for a resource to complete a given task. One person completing a task in one day would be the equivalent of one man-day. It is necessary to calculate man-days and durations for each construction activity to accurately understand the full scope of work and expected costs. Tasking, estimating, and reporting are always done in eight-hour man-day increments to create a standard all bidders adhere to, this way the number of days everyone is using equates to the same number of hours. Since each workday is a standard eight hours, overtime related costs that some contractors may try adding at this stage are minimized.

MDs = QTY OF WORK / UNIT SIZE X MHRS PER UNIT / 8 X DF

Production Efficiency Factors (PEF)

PEFs are the first step in adjusting our man-day estimates based on unique circumstances and/or location. The PEF aims to adjust for elements that make a crew more or less effective. When figuring out a PEF, we look at factors that affect us on the job. The availability factor will cover things that take us away from the work site. The chart below shows the eight production elements in the left column. We'll think about how each of these elements impacts each task for a specific crew, location, equipment state, and so on. We'll give a production score between 25 (Low Output) and 100 (High Output) for each element, with 67 being average. These eight scores are then averaged to get your Production Efficiency Factor (PEF).

	Low Production			Average Production			High Production		
	25	35	45	55	65	75	85	95	
1. Workload	Construction Req High, High Overhead			Construction Req Avg., Avg. Overhead			Construction Req Low, Low Overhead		
2. Site Area	Poor Area, Cramped Working Area			Working Area Limited, Avg. Access			Large Working Area, Ready Access		
3. Labor	Inexperienced Crew, Not Motivated			Adequate Experienced Crew, Motivated			Experienced Crew, Highly Motivated		
4. Supervision	Poorly Trained and Inexperienced			Avg. Trained and Experienced			Highly Trained and Experienced		
5. Job Conditions	Detailed Work Required			Average Quality of Work Required			Rough Work Required, No Detailed Work		
6. Weather	Abnormal Weather: Rain, Heat, Cold			Moderate Weather: Rain, Heat, Cold			Favorable Weather: Rain, Heat, Cold		
7. Equipment	Poorly Maintained Equipment			Fair Condition of Equipment			Excellent Maintained Equipment		
8. Logistical	Slow Procurement and Delivery			Normal Procurement and Delivery			Fast Procurement and Delivery		

Sample Exercise:

Production Element	Percentage	Remarks
1. Workload	67	No specific impact
2. Site Area	75	Good access to work area
3. Labor	35	Crew inexperienced, a lot of OJT required
4. Supervision	75	Experienced supervision
5. Job Conditions	45	Detailed work required
6. Weather	67	No specific impact
7. Equipment	70	Sufficient tools in good condition
8. Logistical	75	Materials on hand in good condition
TOTAL:	**509**	

PEF = 509 (Total) / 8 (# of Categories) = 63.6

Delay Factor (DF)

Before we can adjust our man-day estimates, we must convert our production efficiency factor into a delay factor. We can find the delay factor by dividing the acceptable average of 67 by the PEF (DF=67/63.6=1.05) or by using the table below:

Production Efficiency Versus Delay Factor

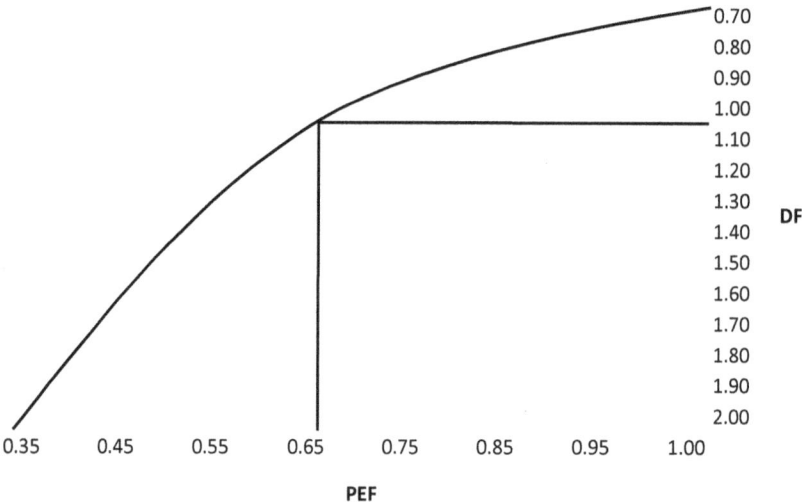

DF (vertical axis, top to bottom): 0.70, 0.80, 0.90, 1.00, 1.10, 1.20, 1.30, 1.40, 1.50, 1.60, 1.70, 1.80, 1.90, 2.00

PEF (horizontal axis): 0.35, 0.45, 0.55, 0.65, 0.75, 0.85, 0.95, 1.00

Applying our 1.05 or 5% delay factor, we can now modify our initial man-day estimate. This method has its limits. For instance, if we're working outdoors in severe weather, with all other factors deemed average, we might get a PEF of 62 and a delay factor of 1.08. This 8% rise in our man-day estimate would not sufficiently account for operating in extreme weather conditions. As with anything, use common sense when calculating the PEF. Come up with a justifiable PEF and DF and discuss it with others. You should use your experience in determining the correct delay factor to use under each circumstance.

Direct Labor (DL) & Availability Factor (AF)

The Availability Factor takes into account that labor assigned to a project is not available 100% of the time. Availability factors, for planning purposes, are determined by site efficiency factors. They vary from **.85 for an average** and anything above or below that average is above or below average efficiency. An availability of .75 means the crew is available for 75% of an eight-hour workday, which is lower than average and should be taken into consideration. Considerations include break times, lunch time, transportation, security screenings, etc.

On-site labor man-days are not as easy to calculate for several reasons. First, unlike office activities, a typical shift in construction is considered a workday (WD) of 10 to 12 hours, so we need to normalize the data. To determine this, consider how much of a ten-hour workday is actually used conducting direct labor. If you determine your crew actually spends eight hours a day on the jobsite conducting work directly related to work-in-place, then your availability factor would be .80. This factor combined with other direct labor factors helps the project manager develop the man-day capability (MC). Man-day capability is the number of direct labor actually used for direct labor on the jobsite and is critical to arriving at accurate man-day estimates. The formula is below:

$$MC = DLxWDxAF$$

DL = Direct Labor WD = # of hours at work AF = Availability Factor

The MC equation listed above is also used to determine construction activity duration by replacing MC with the estimated MD, inputting crew size (CS) in place of direct labor assigned (DL), and applying the availability factor (AF) and then solving the number of workdays required, otherwise known as the duration. The duration equation is below:

Duration = MD estimate / CS / AF

The activity duration is increased by including the availability factor to account for time lost from the project site. The actual crew we would expect to see on the jobsite on the average day would be the assigned crew divided by the availability factor.

SUBCONTRACTORS

Subcontractors are an integral part of any project and the selection process is important for multiple reasons. Of most concern to General Contractors (GC) is the price of the work, the reliability of the subcontractor, and their quality of work that becomes a reflection on you and the project as a whole. Many base their selection on price alone, assuming the cheapest option will be the most profitable. And while cost is a high priority we always advise looking at several other factors such as your past experience and relationship with the subcontractor, their overall track record for success, their geographic proximity to the jobsite where the work will take place, and their overall safety rating. In the past, the federal procurement process only looked at price but because of everything we just mentioned they now use a weighted formula, known as Lowest Price Technically Acceptable (LPTA) or Best Value, where all these considerations are accounted for.

You will generally have anywhere from several weeks to a month to get your estimates together, which allows plenty of time to secure bids from your subcontractors. RFIs will delay the process even more, so there shouldn't be any pressure to rush this part of the process. Take the time to thoroughly review all bids and confirm that their quantities and dollar amounts all match what was on the drawings and specifications. You never want to find yourself in a position later where you are awarded the job only to find out your subcontractors bid incorrectly and you must now justify change orders to the Client.

TOOLS AND EQUIPMENT

The Project Manager must have access to the list of tools and equipment available to him at any site. The tools must be available to support the method of construction used in planning the project. Production rates and construction schedules will be affected if alternate equipment must be used. If the tools are not available, rental of tools is always an option, provided the budget allows and rental equipment is available. Remember, if you anticipate the need for rental tools, those bidding costs must be figured into the budget for that particular job.

REVIEW OF BID PACKAGES

Every Company will have their own procedures and protocols for what they want to see in the bid packages they receive, so as a potential contractor for a job you should be prepared to comply with exactly what they are looking for. The best way to do that is always read the description of what is to be included in your submission and follow all the instructions exactly as they were written. Below is a list of best practices for reviewing your submissions to make sure you have addressed the most common requests:

1. Determine where to locate and obtain the bid forms and be aware of submission deadlines and the acceptable forms of delivery (email, overnight mail, hand delivery, etc.)
2. Note who to contact with any questions prior to submission. Do not assume you know the answer if something is not crystal clear.
3. Follow the specific instructions on what fields on the form need to be filled out, what forms can be left blank, and what modifications, if any, can be made to the pre-existing fields.
4. Determine if the job requires prevailing wages or not before calculating labor costs.
5. Determine if material Supplier Authorizations are required with the bid.
6. Be sure you are able to start on the scheduled dates if awarded the job.
7. Know where to locate the plans and drawings.
8. Know how much the bid bond or deposit amount will be required and ensure you are able to provide it by the specified date.

9. Know the types of bonds required once the job is awarded and be sure you can comply.
10. Know when the deadline to submit substitution (or equal) requests for any items differing from the original specifications.
11. Be aware of the number of alternates.
12. Know the retainage amounts or percentages.
13. Be aware of the liquidated damage penalties in both dollars and days.

PRE-BID MEETING AGENDA

Below is a sample of a Pre-Bid Meeting Agenda:

1. **Welcome**

2. **Project Team Introduction:**
   a. Owner
   b. Design/Builder
   c. Architect
   d. Construction Superintendent

3. **Project Summary**

4. **Bidding Instructions**
   a. Bidding Information
   b. Submit Bids to (NAME and DATE)
   c. Bonding – AIA A312 Performance and Payment Bond
   d. Bid Sheets
   e. Worker Compensation

5. **Summary of Work**

6. **Review of Special Tasks for Individual Contractors**
   a. Payment Procedures
   b. Submittals and Shop Drawings
   c. Bid the Schedule
   d. Quality Assurance Testing Requirements
   e. Temporary Facilities and Controls

7. **Review of Special Tasks for Individual Contractors**
   a. Substitutions

b. Project Close Out
    i. Spare Parts
    ii. Operation & Maintenance Manuals
    iii. As-Built Drawings
    iv. Warranty
    v. Sample Contract
    vi. Safety Procedures

8. **Acknowledgment of Amendments**: It is common for the RFP contract documents to undergo multiple revisions. They can range from extending the bid date to changing the scope of work. This step lets the other party know you have accounted for all changes in your bid.

9. **Significant or Unusual Risk:** Prior to submitting a proposal, signing a contract, or performing work involving any of the following risks or exposures, the proposal and contract must be reviewed and approved by a senior member of your organization.

10. **Risk Contracts:** Any contract for fixed-price, lump sum, or guaranteed maximum price work that fails to contain any of the provisions required for such work

11. **Indemnification or Insurance for Others' Negligence:** Any contract that requires the Company to indemnify, ensure, or protect a Customer or other party from that party's own negligence

12. **Warranty**: Any contract that does not have defined, exclusive remedies for breach of warranty restricted by a 1 year period of time

13. **Limit of Liability:** Any contract that fails to contain an absolute and legally binding limit on the Company's overall liability to the Customer or that contains a limit on an amount exceeding $1 million and three times the contract value. The limit of such liability shall not exceed $5 million, however, without further approval by senior management. Limitations of liability may exclude third party claims for bodily injury or property damage to the extent that such injury or damage arises out of the Company's negligence.

RISK ASSESSMENTS

Risk is a very real factor your estimating team needs to take into account when preparing to bid on a job. Insurance is not cheap, and the general liability you carry is only the tip of the iceberg. Each project you bid on will require various types of surety bonds, and each will come at an additional cost. Also, depending on your Company's size and track record, you might not always qualify for the highest limits needed, which would make bidding on jobs with those requirements a waste of your time.

It is also important to draw distinctions between single-project bonding limits and aggregate limits, which apply to all projects you have underway. Just because you have a single bond limit that will cover a job does not mean you can bid a job exceeding your aggregate limit. For example, if you have a single limit of $5 million and an aggregate of $10 million, you will not be able to bid a job with a $4 million budget if you already have projects underway using more than $6 million of your aggregate total. This is part of the reason that all the assessments in this section are critical to your bidding success.

General Liability Insurance

This is the foundational insurance every General Contractor needs to have before thinking about bidding or working on any job. Its purpose is to make sure any damage or injuries occurring to material or personnel on a job site are covered and will not become a financial burden on the Project Owner. The General Contractor shall maintain the insurance coverage limits specified within each individual job proposal and furnish the Owner with certificates evidencing the insurance. Such certificates shall provide for at least 30 days prior written notice to the Owner of any policy cancellation.

The general insurance requirements for a General Contractor regardless of the job is generally as follows:

- Worker's Compensation and Employers Liability Insurance: Required under laws applicable to the work, which shall cover General Contractor's employees engaged in the work.
- Comprehensive General Liability Insurance (including blanket contractual liability coverage) and Automobile Liability Insurance: Covering all owned and non-owned, registered automobiles or trucks used by or on behalf of General Contractor, with recommended coverage amounts of combined single limits for bodily injury and/or property damage in the amount of $1 million and $1 million aggregate.
- General Contractor's insurance coverage shall not limit the General Contractor's obligations hereunder, nor shall such insurance be construed to increase any limitations on General Contractor's liability otherwise provided in this proposal.

Bid Bond

A bid bond is going to ensure any mistakes made in the estimating process are not passed along to the Customer. This is important because errors in the bid that lead to your offer becoming impossible to meet can leave everyone in a difficult position. When your bid does not leave a profit margin, there would be no way for your Company to complete the job at the agreed upon price. This would leave the Customer in a position where they need to award the job to the second lowest bidder, which could be at a significant increase over your bid and leaving the Customer in an unprofitable situation. The bid bond will pay the Customer the difference between your bid and the next best bid.

Builder's Risk Policy

Builder's risk is more of a traditional insurance policy designed for when the building is under construction. It covers the cost of stolen material (copper, aluminum, or other valuable components) and also damage to the project as well. Your general liability is more for injuries and will likely not cover most of the items listed above. See Appendix 2-1 for an overview of what is required on a standard questionnaire.

Performance Bond

A performance bond kicks in if you cannot finish the job for any reason, the most common being that your Company has run out of money or you were fired for gross negligence or other serious factors where the Customer sees it more prudent to start over with someone else. The bond pays the costs of bringing in a new contractor to fulfill the remainder of the duties you left incomplete. A sample can be found in Appendix 2-2.

Payment Bond

A payment bond insures you are paying subcontractors and suppliers. As you invoice the Customer for project milestones, they will pay you the lump sums as specified and agreed. The work completed to meet these milestones is performed by multiple contractors under your supervision and it is your responsibility to make sure the funds are disseminated to the right people in the right amounts. If you fail to meet that obligation, the payment bond will take effect. A sample can be found in Appendix 2-2.

ADDITIONAL CONSIDERATIONS

Depending on the size, scope, and geographic location of your project, there are other variable and fixed costs to factor into the total overhead of a project. From a facilities standpoint, your contractors will likely need temporary toilets, construction trailers, office equipment, and other forms of business services such as phone and internet. From a personnel standpoint, unless you have the in-house resources on staff, it is advisable to include the costs for project managers, project administrators, file clerks, and any other relevant specialization you are lacking. It is always better to err on the side of caution by including these costs and not needing them then omitting them and risk losing money by the end of the project.

For larger jobs, you might find yourself in need of more specialized roles and equipment that would not otherwise be a part of a standard project. At ACE, we pride ourselves on providing Quality Control, Safety, and Management personnel to bridge any gaps in your staffing. Vehicles and Conex boxes will also start coming into play and it is always advisable to have both an idea of what the costs will be as well as the availability so that you do not wind up delaying a job before it has ever started.

CONSTRUCTION ACTIVITY SUMMARY SHEET (CASS)

We have referred to the CASS throughout this chapter, but up until now we have not had enough data to sufficiently complete it. In addition to activity description and scheduled dates, all required resources are shown, along with safety, quality, and environmental requirements. In addition there is a comments section for any additional information relevant to a particular activity. A sample CASS is provided in Appendix 2-3.

PROPOSAL PROCEDURES

The proposal procedures are established to prevent confusion, request a proposal number, and follow the outlined proposal process. Nothing should be sent (fees, quotes, etc.) without having acquired a proposal number. The following standard process should be followed upon receipt of either the scope of work (SOW) or notification from the Client.

1. Project Manager receives Scope of Work from Client.
2. Project Manager gives Scope of Work to Project Administration to be backed-up.
3. Project Administration gets Proposal Request Form and completes basic information; copies SOW to Contract Administrator.
4. Project Administration forwards Proposal Request Form to Project Manager for proposal specific information.

5. Project Manager forwards completed copy of Proposal Request Form to Business Development (BD). Proposal # and Internal Order # are generated for the file. Then forward a copy of the Proposal Request form to Project Controls, Project Administration, and Project Manager. BD will update the Proposal Log.
6. Project Manager completes the proposal with assistance from Project Administration.
7. Project Administration forwards the proposal to Client, Marketing, and Project Manager.
8. Project Administration is to receive ANY & ALL correspondence relative to this proposal; Project Administration will copy ALL correspondence to the Business Development Administrator.
9. Project Manager notifies Project Administration upon notification of status of proposal (award, loss, etc.) and forwards contract documents to Project Admin.
10. Project Administration copies contract documents to BD.

A project number is to be set up for all awarded/approved projects. Unless approved by a project executive, you should not be doing project work under a proposal number. These procedures apply to all proposals submitted to current or potential Clients where contract terms and conditions are not established. Where terms are established, the procedure applies to opportunities where terms are different from existing no-risk blanket agreements or opportunities where any risk will be taken.

Prior to the proposal review meeting, the assigned proposal manager will complete a Proposal Review Form (Appendix 2-4). The proposal review meeting will be attended by the Operations and Business Development personnel.

Proposal Clauses

The proposal, contract, or intention to perform work must be reviewed and approved by a senior official at least two weeks before submitting, signing, or performing any actual work if it involves any of the following risks or exposures:

- **Consequential Damages**: Any contract that fails to contain an absolute and legally binding exclusion of liability for indirect, incidental, or consequential damages, including, if applicable, from multiple Owners. If the Company is a contractor, the Owner and all intermediate contractors must also agree to release the Company from these risks.

- **Limit of Liability**: A maximum amount should be stated (i.e., $5 million) in all contracts. Any contract that fails to contain an absolute and legally bonding limit on the General Contractor's overall liability to the Customer shall be capped at $10 million. Limitations of liability may exclude third party claims for bodily injury or property damage to the extent such injury or property damage arises out of the Company's negligence. Notwithstanding anything to the contrary, the remedies in a proposal are exclusive and in no event shall the liability of the General Contractor, its subconsultants, contractors, suppliers, agents, insurers, affiliates, or employees, to Owner for any and all damages, claims, losses, costs or expenses (damages) exceed in the aggregate an amount equal to the compensation paid for the services provided under this proposal.

- **Uninsured Risk**: Any contract exposing the Company to uninsured liability for personal injury or for damage to the property of others (e.g., for property within the Company's care, custody and control).

- **Indemnity:** The General Contractor shall indemnify, hold harmless and defend Owner and its officers, directors, employees, and agents (collectively referred to as "Owner") from and against any and all damages, claims, losses, expenses and liabilities, including attorney's fees, claimed by third parties for property damage and bodily injury, including death (collectively referred to as "Claims"), sustained by Owner to the extent such Claims are caused by the negligence of the General Contractor. Any defense costs and attorney's fees incurred in connection with the foregoing indemnity and defense shall be borne by the Owner to the extent of its negligence.
- **Non-Competition, Non-Solicitation, or Non-Hire Agreements**: Any contract, letter, or other agreement that limits or restricts the Company from offering services to, or competing or doing business with, any other Company, or from hiring or offering to hire any person.

Legal Department in the Proposal Process

The Legal Department must be consulted in order to determine whether releases of consequential damages, or limitations of liability, if granted by a Customer, are legally binding. In addition, all contracts involving fixed-price, lump sum, or guaranteed maximum price work, or any of the significant or unusual risks described, must be reviewed by the Legal Department, in advance of their being signed.

If a Company provides work in an industry where certain terms and conditions are not acceptable to Customers (e.g., waivers of consequential damages or limit of liability), or the Company engages in a scope of work that may mitigate certain risks or exposures, the

Company may seek a "blanket waiver," or permission to perform the work without obtaining specific approvals for each contract. The Company executive shall initiate the written blanket waiver request for review by a lawyer. Approved blanket waivers must be signed and in writing. All original approved blanket waivers shall be archived in the Legal Department.

Significant or Unusual Risk Approval Matrix

Description of Proposal, Contract	Value of Work or Services	Required Approvals
Any Fixed Price, Lump Sum or Guaranteed Maximum Price Contracts Without Appropriate, Force Majeure, Changes and Termination Clauses	All	VP Level
Indemnification or Insurance for Others' Negligence	< $5,000,000	VP Level
	>$5,000,000	President
A Warranty Provision that Fails to Contain Defined, Exclusive Remedies Restricted by a Time Period of Two Years or Less	All	VP Level
A Warranty Provision that Fails to Contain Defined, Exclusive Remedies Restricted by any Time Period, or Contains such a Restriction but the Time Period is in Excess of Two Years	All	President
Limit of Liability	$1,000,000 to $5,000,00 and Three Times the Contract Value	VP Level
	$5,000,000 to 10,000,000	President
	> 10,000,000	President

Uninsured Risk For Personal Injury or Damage to Property of Others	All	VP Level
Liability for Environmental Services or Professional Opinions	All	VP Level and President
Retention of International Representatives	All	VP Level and President
Any Contract that Fails to Contain a Waiver of Consequential Damages	All	VP Level and President
Non-Competition, Non-Solicitation, or Non-Hire Agreements	All	VP Level and President

Key Notes on Proposal

REQUIRED USE: An executive summary shall be generated in advance of the proposal review process so all issues can be resolved and incorporated into the final document prior to the formal proposal review meeting. The completion of an executive summary does not eliminate the need to fill out the other required documents as specified.

RESPONSIBILITY: The proposal manager shall be responsible for producing and submitting this document with the assistance of operations, BD, and/or finance, as required.

APPLICATION: An executive summary is required for all proposals requiring executive level approval and above as defined on the approval matrix of this procedure as shown above.

OUTLINE OF EXECUTIVE SUMMARY TEMPLATE: (Also see the sample Executive Summary in Appendix 2-5)

1) **Scope Section:**
    a) Scope of work: Provide an overview of the project to include the technology, location, major equipment suppliers, field conditions, etc.
    b) Scope of services: Define summary level services (i.e., engineering, construction, maintenance, lease, etc.).
    c) Scope-related risk issues and/or unknowns: Define all issues that would be potential risk (undefined scope areas, Client performance of work, unknown site conditions, etc.).

2) **Schedule Section:**
    a) Milestone dates: Define the Award date, start work, completion, startup, plus any key milestones tied to fee/risk.
    b) Analysis: Provide analysis of the total schedule cycle time versus similar work and likelihood of making the target dates.

3) **Pricing Section:**
    a) Summary by major cost element: provide a breakdown of the total cost to include the following minimum items:

Cost Element	Amount($M)	% of Total Cost	Responsibility (Contractor/Client/Subs)
Equipment			
Field material			
Field labor			
Subcontracts			
Engineering			
Contingency			
Fee			
Other			
Total			

    b) Estimate basis: Define the basis for the estimated values as well as rationalizing the proposal cost against the total net contribution expected as follows:

        i) Define markups/multipliers versus break even.

        ii) Define basis for equipment/material pricing (hard quotes, estimates, etc.).

        iii) Define risks in the estimate (factors, allowances, subcontractor bid quality).

        iv) Define proposal cost and net value (deduct billable labor).

    c) Cash flow projection: Chart the cash flow throughout the project including as a minimum payment schedule, spendout plan, payment terms, retainage.

4) Risk/Opportunity Analysis – by major cost element:

    a) Provide narrative of each section describing any risk associated and/or opportunity.

5) **Contract Section:**
   a) Terms & Conditions: Define exceptions and quantify risk.
   b) Specific T&C analysis: Address the following:
      i) Warranty and correction
      ii) Indemnification
      iii) Consequential damages
      iv) Insurance coverage opposite
      v) Liquidated damages

CONTRACTS AND CONTRACT CHANGE

Typically, the General Contractor's contract specialist prepares contracts and changes to those contracts. Contracts are distributed to subcontractors and suppliers. These items are recorded by the project assistant on the contract log when they are sent out from the General Contractor. The contract statuses are tracked via logs prepared by the project administration and are issued to the Project Manager weekly. General Contractors are not allowed to begin work until both parties sign a fully executed contract. Any changes to contracts should follow the procedures outlined in Chapter 8 on Project Execution.

CHAPTER 3
ENVIRONMENTAL

Environmental protection and hazardous waste disposal are two serious concerns in today's business world. Hazardous substances like solvents, corrosives, adhesives, protective coatings, and various finishes are commonly found at construction sites. As the Project Manager, ensuring worker safety and environmental protection falls under your direct responsibility. You must safeguard both your team members and the surrounding environment. Strict regulations and monetary penalties exist for environmental violations, affecting all companies regardless of their operational location. The consequences of failing to maintain an environmentally compliant worksite can include monetary penalties and even imprisonment. While you aren't expected to be an environmental expert, you have support available. Contact your organization's environmental coordinator or safety officer right away if you encounter any environmental concerns (chemical releases, regulatory documentation, strategic preparation, etc.).

Environmental and hazardous waste specialists often face criticism for seemingly hindering operations and development. However, with effective coordination and proper preparation, environmental challenges can be addressed through straightforward procedures. Success lies in seeking guidance before issues escalate beyond control. Laws and rules are in place for a reason, and that usually stems back to a precedent where there was a negative impact. As long as everyone is doing their fair share by following the guidelines in this chapter, it will make for a smoother and safer construction project.

Environmental Protection Plans (EPP), along with other critical submittals, are discussed in greater detail in other parts of this book.

DEFINITIONS

The following terms are often confused in discussing environmental issues:

- **Hazardous Waste:** A hazardous material or substance which is to be removed or being removed from a construction site and is no longer required in the construction process.
- **Acute Hazardous Waste:** Can cause death, debilitating personal injury, or serious illness in individuals who are exposed.
- **Non-acute Hazardous Waste:** Less dangerous than acute waste, but still a liability that should be mitigated for on the job site.
- **Hazardous Material:** Unlike hazardous waste, hazardous materials are required for future construction work. The rules governing the storage, handling, and use of hazardous materials are covered in Chapter 10 of this handbook.

- **Hazardous Waste Minimization:** The process of limiting hazardous waste to the greatest extent possible. The best answer to hazardous waste disposal is to not have any, if possible.

OBTAINING PERMITS

When we get the drawings from the designer of record, it shows the layout of the building site and where the water run-off is supposed to go. From an environmental standpoint, this water control issue is one of the major permits you will need to secure. In practice, 99% of the time your designer will handle this for you, but it is still worth noting due to all the federal, state, and local regulations. Any project in the United States that disturbs more than one acre of land requires a NPDES permit, short for National Pollutant Discharge Elimination System. The permit is issued by the Environmental Protection Agency to control the amount of silt and run-off from your project, ensuring construction debris and harmful pollutants are not making their way into local waterways. Your permits will outline your best management practices (BMP), which will need to be followed throughout the construction process. Once these permits are obtained, they must be displayed on-site for the duration of the project.

EROSION CONTROL AND WATER RUN OFF

Once your permits have been issued in accordance with the BMPs, you need to begin implementing the various methods needed to stem the flow of water runoff and control erosion. Depending on the conditions and complexity of the job site you might find yourself in need of one, some or all of the following management strategies:

Silt Fences

These are impermeable nylon fences trenched roughly six inches into the ground and secured with wooden stakes. Fill is then backed up to it, creating a barrier to filter water. The nylon catches any sediment and silt to prevent it from winding up in public water sources. They are most effective when installed at the low points of the site.

Check Dams

Composed of stone or rock piles, check dams are used as an obstacle to slow the flow of water and make it less likely to grab sediment as it runs its course. This is the same theory behind why landscapers use river rock or other smaller stones to filter water away from houses so they do not absorb excess water and flood.

Temporary Seeding

Grass does a great job at slowing the flow of water. As jobsites are prepared, the leveling process removes all existing greenery and leaves only a layer of soil. As a rule of thumb, if work is not scheduled on a particular section of the jobsite for two weeks or more, temporary seeding needs to be laid down to assist with erosion control.

Erosion Control Matting

Generally, this is nylon netting filled with straw bales, hay, or any other type of fibrous material that gets laid down to take the place of temporary grass while it grows to a sufficient level.

WASTE MANAGEMENT

Plans for how you are going to deal with waste management issues at the outset of a project allow for an easier set up of the jobsite and a higher likelihood of adhering to all federal, state, and local regulations. Recycling is a big topic and varies widely even on a local level, so it is always best to plan for the most stringent measures possible. Many contractors opt to have separate dumpsters to sort wood, metal, and trash while still on-site to allow for quicker and easier removal to the appropriate facility. Then there are the concerns of where to store and move contaminated soil or other toxic elements like asbestos and lead.

POTABLE AND NON-POTABLE WATER

While potable water is more of a safety concern than it is environmental, it is still important to mention both here. First, both types of water should be clearly labeled to ensure the wrong type of water is not used in the wrong situation. While using potable water for non-potable activities will not lead to a safety or environmental issue, it could leave you with a shortage of drinking water for the workers. Having a sufficient amount of non-potable water is crucial for performing specific construction related activities that could otherwise lead to hazardous conditions. Dust control from the moving of sediment and crushing of aggregate stone keeps the particles out of the air, a direct benefit to the health of your workers and native wildlife.

INSPECTIONS

Inspections are critical to maintaining the integrity of your environmental protection efforts. It is an expectation for the contractor to adhere to certain minimum standards, generally specified in your permits, because the local enforcement agencies will be there to keep you honest. BMPs will dictate the specifics, but expect an inspection of your erosion control measures within a certain number of days after significant rainfall and at routine intervals under normal conditions. These inspections will include ensuring silt fences are not torn, check dams are still in place, and all temporary seeding or erosion mats are in place.

Outside of your obligation to preserve the environment you work in, the penalties for failing to do so can become substantial. For initial infractions, inspectors will allow a requisite amount of time to correct the issues. Repeat offenses will result in monetary fines, as specified in the permits, and can total tens of thousands of dollars. In addition to inspecting your controls, they will also trace the downstream water where the runoff travels and test for turbidity levels. If found to be in excess of acceptable levels, this needs to be corrected as well. As an ultimate penalty, jobsites can be shut down for continued negligence.

HAZARDOUS WASTE PLANNING

During the project planning phase, you must consider the effects of hazardous materials and hazardous waste on your project and schedule. The CASS should identify activities that involve the use of these items. The first step in any planning process should revolve around the jobsite storage of hazardous materials until they are needed and hazardous waste until it can be removed. These storage and personal protective measures, along with Safety Data Sheet (SDS) requirements, are listed on the SDS for the product you are using (SDS was previously referred to as MSDS - Material Safety Data Sheet - but is now shortened to SDS). Plan for delivery of proper storage equipment prior to having hazardous materials delivered to the jobsite.

There are several best practices for dealing with hazardous materials and waste:

- Minimize the quantity of hazardous material needed on the job site by only storing enough material for two weeks' worth of work at a time, if feasible.
- Remove hazardous waste as soon as possible and contain it at all times.
- Proper labeling of hazardous materials is critical. Properly labeled waste can be disposed of relatively cheaply. Unidentified waste that must first be analyzed, then disposed of properly, is more costly. The best policy is to identify waste and dispose of it properly.
- Avoid mixing unlike types of waste. Some material can be recycled. This can only happen if materials are not mixed. The best method for disposal is to label it and dispose of it properly.
- Have a plan to keep your crew informed of the types of materials on the job. Each crew member has the right to know the effects of working with hazardous materials. This information is found on the SDS. The "Right To Know" policy is a law that protects employees' rights. A clear defiance of this law is a crime.

SPILL RESPONSE

The protocol for managing a spill follows two key principles: contain and report. While it may seem straightforward, your initial action must be to stop the spill from expanding and secure the affected area. Shut down the source. If the liquid is not moving, do not disturb it. Never try to conceal or cover up even minor spills. This will only complicate the investigation and cleanup process. Once containment is achieved, notify your supervisors promptly. Be prepared to share details about the substance involved, the quantity spilled, and when the incident occurred.

CHAPTER 4
CONSTRUCTION SCHEDULING

To finish projects on time, we need to create realistic, doable schedules during planning. Key to this is putting tasks in the right order and giving them realistic timeframes. By mapping out steps forward and backward, we can find our critical path. This path, the longest route through a project, shows us key checkpoints we must hit to stay on track and finish when planned. Without a good schedule and critical path, we risk delays and missed deadlines. Careful planning upfront helps ensure smooth execution and timely completion of construction projects.

Since we are going to be looking at Level IIs and Level IIIs throughout this chapter, let's start with the broad scope of what we consider a Level I activity for the sake of this book. In an individual project, a Level I would describe the project as a whole, but we are looking at this in terms of your Company's entire portfolio of projects in progress. Whether you have three, five, or a hundred projects, Level I encompasses them all. As the levels increase, the activities become more detailed and complex — for example, a Level I might be "Construct New Hospital Wing," while a Level III could break that down into "Construct New Hospital Wing > Preconstruction > Long Lead Items."

CRITICAL SUBMITTALS

Even after you have been awarded the contract for a project, there are still a number of items you must provide to the Owner before you can even think about mobilizing labor or material to a job site. We discuss these at a high level here, as they are covered in more detail in other areas of the book. It is essential that each of these submittals is accurately captured and reflected in the project schedule to ensure timely approval and avoid delays. In no particular order, these critical submittals are:

Accident Prevention Plan (APP)

The following guidance on Accident Prevention Plans (APP) and site safety procedures reflects commonly accepted practices derived from industry-standard construction specifications and historical editions of the U.S. Army Corps of Engineers EM 385-1-1 Safety Manual. While these general principles are broadly applicable, always consult your project's specific contract documents for definitive safety requirements.

For every building operation, all identified risks and remedial measures will be documented on the CASS. The project safety plan (included in this chapter) is subsequently developed detailing the risks and remedial actions from the CASS. A summary document outlines the safety protocol detailing necessary training requirements and essential equipment specifications. The project safety plan must be posted on the jobsite. A daily jobsite safety inspection (Appendix 4-1) will be performed daily by the Project Superintendent.

The General Contractor shall use a qualified person to prepare the written site-specific APP in accordance with the format and requirements of USACE EM 385-1-1. Minimum basic requirements from the EM 385-1-1 are listed herein. All paragraph and subparagraph elements in USACE EM 385-1-1, Appendix A, "Minimum Basic Outline for Preparation of Accident Prevention Plan (APP)" should be covered. Where a paragraph or subparagraph element is not applicable to the work to be performed, indicate "Not Applicable" next to the heading.

The APP shall be job-specific and shall address any unusual or unique aspects of the project or activity for which it is written. The APP shall interface with the General Contractor's overall safety and health program. Any portions of the General Contractor's overall safety and health program referenced in the APP shall be included in the applicable APP element and made site-specific. The Government considers the General Contractor to be the "controlling authority" for all work site safety and health of the subcontractors. General Contractors are responsible for informing their subcontractors of the safety provisions under the terms of the contract and the penalties for noncompliance, coordinating the work to prevent one craft from interfering with or creating hazardous working conditions for other crafts, and inspecting subcontractor operations to ensure accident prevention responsibilities are being carried out.

The APP shall be signed by the person and firm (senior person) preparing the APP, the General Contractor, the on-site superintendent, the designated site safety and health officer and any designated CSP and/or CIH and submitted to the Contracting Officer, fifteen calendar days prior to the date of the preconstruction conference for acceptance. Work cannot proceed without an accepted APP.

The Contracting Officer reviews and comments on the General Contractor's submitted APP and accepts it when it meets the requirements of the contract provisions. Once accepted by the Contracting Officer, the APP and attachments will be enforced as part of the contract. Disregarding the provisions of this contract or the accepted APP will be cause for stopping of work, at the discretion of the Contracting Officer, until the matter has been rectified.

Once work begins, changes to the accepted APP shall be made with the knowledge and concurrence of the Contracting Officer, project superintendent, SSHO, and Quality Control Manager. Should any unforeseen hazard become evident during the performance of work, the project superintendent shall inform the Contracting Officer, both verbally and in writing, for resolution as soon as possible. In the interim, all necessary action shall be taken by the General Contractor to restore and maintain safe working conditions in order to safeguard on-site personnel, visitors, the public, and the environment.

Copies of the accepted plan will be maintained at the resident engineer's office and at the job site. The APP shall be continuously reviewed and amended, as necessary, throughout the life of the contract. Unusual or high-hazard activities not identified in the original APP shall be incorporated in the plan as they are discovered. You should always reference specific requirements in your Project Specifications or the EM385, but minimum requirements for your APP should include, at minimum:

1) Names and qualifications of all site safety and health personnel designated to perform work on this project
2) Qualifications of competent and of qualified persons
3) Confined Space Entry Plan

4) Health Hazard Control Program

5) Crane Critical Lift Plan

6) Alcohol and Drug Abuse Plan

7) Fall Protection and Prevention (FP&P) Plan

8) Training Records and Requirements

9) Occupant Protection Plan

10) Lead Compliance Plan

11) Asbestos Hazard Abatement Plan

12) Site Safety and Health Plan

13) Polychlorinated Biphenyls (PCB) Plan

14) Site Demolition Plan

15) Excavation Plan

16) Crane Work Plan

Activity Hazard Analysis (AHA)

The following section on Activity Hazard Analysis (AHA) outlines widely recognized safety planning practices informed by the U.S. Army Corps of Engineers EM 385-1-1 standards. Always refer to current contract specifications and site-specific safety protocols for detailed requirements applicable to your project.

The APP shall also contain a section specific to Activity Hazard Analysis (AHA) in accordance with USACE EM 385-1-1. The AHA must be submitted for review at least fifteen calendar days prior to the start of each phase for every operation involving a type of work presenting hazards not experienced in previous project operations or where a new work crew or subcontractor is to perform work. The analysis must identify and evaluate hazards and outline the proposed methods and techniques for the safe completion of each phase of work.

The AHA list will be reviewed periodically (at least monthly) at the General Contractor supervisory safety meeting and updated as necessary when procedures, scheduling, or hazards change. Activity hazard analyses shall be updated as necessary to provide an effective response to changing work conditions and activities. The on-site superintendent, site safety and health officer, and competent persons used to develop the AHAs, including updates, shall sign and date the AHAs before they are implemented. A sample AHA can be found in Appendix 4-2.

Environmental Protection Plan (EPP)

The information provided in this section regarding the Environmental Protection Plan (EPP) reflects practices commonly found in industry-standard construction specifications. For project-specific requirements, consult your contract documents or contact the local contracting office for guidance. Additionally, ensure your project complies with all applicable state and local environmental regulations.

The EPP shall include any hazardous materials (HM) planned for use on the job site and included in the HM Tracking Program maintained by the Safety Department. To assist this effort, submit a list (including quantities) of HM to be brought to the job site and copies of the corresponding SDS (Safety Data Sheet) to the Contracting Officer. At project completion, remove any hazardous material brought onto the job site. Also account for the quantity of HM brought to the job site, the quantity used or expended during the job, and the leftover quantity which:

1) may have additional useful life as a HM and shall be removed by the General Contractor, or

2) may be a hazardous waste, which shall then be removed as specified herein.

The EPP shall list and quantify any Hazardous Waste (HW) to be generated during the project. In accordance with project and base regulations, store HW near the point of generation up to a total quantity of one quart of acute hazardous waste or fifty-five gallons of non-acute hazardous waste. Move any volume exceeding these quantities to a HW permitted area within three days. Prior to the generation of HW, contact the Contracting Officer for labeling requirements for storage of hazardous wastes. Also plan to substitute materials as necessary to reduce the generation of HW and include a statement to that effect in the Environmental Plan.

Always contact the Contracting Officer for conditions in the area of the project that may be subject to special environmental procedures and include this information in the Preconstruction Survey. Describe in the EPP any permits required prior to working the area, and contingency plans in case an unexpected environmental condition is discovered. Obtain permits for handling HW, and deliver completed documents to the Contracting Officer for review. File the documents with the appropriate agency, and complete disposal with the approval of the Contracting Officer. Deliver correspondence with the state concerning the environmental permits and completed permits to the Contracting Officer. A template for the acceptable format of an EPP can be found in Appendix 4-3.

Stormwater Pollution Prevention Plan (SWPPP)

This section outlines common components of a Stormwater Pollution Prevention Plan (SWPPP) based on Environmental Protection Agency (EPA) guidance. To ensure full compliance, review the specific state and local stormwater regulations applicable to your project.

The SWPPP illustrates a contractor's plans for sediment and erosion control and are generally required if a project is expected to disturb an acre or more of land, or if the disturbance is part of a larger overall plan. These documents are site-specific and identify potential sources of stormwater pollution. SWPPPs include a description of construction activity, erosion control measures (permanent and temporary), stormwater management, and best practices for contractors as they navigate their project. SWPPPs are beneficial as they can help identify erosion and stormwater concerns before construction begins and avoid unexpected setbacks or environmental issues. Effective stormwater management provides protection of waterways and wetlands, improved quality of waterbodies, conservation of water resources, protection of public health, and flood control. A detailed example of what should be included in a SWPPP can be found in the Appendix 4-4.

Quality Control Plan (QCP)

The information provided in this section regarding the Quality Control Plan (QCP) reflects practices commonly found in industry-standard construction specifications. For project-specific requirements, consult your contract documents or contact the local contracting office for guidance.

The General Contractor shall provide a QCP submitted in a three-ring binder with pages numbered sequentially that covers both on-site and off-site work for approval by the Contracting Officer, including the following in this specific order (Appendix 4-5):

1) QC ORGANIZATION
    a) Names and Qualifications
    b) Duties, Responsibility, and Authority of QC Personnel
    c) Outside Organizations
    d) Appointment Letters
    e) Submittal Procedures and Initial Submittal Register
    f) Testing Laboratory Information
    g) Testing Plan and Log
    h) Procedures to Complete Rework Items
    i) Documentation Procedures
    j) List of Definable Features
    k) Procedures for Performing the Three Phases of Control
    l) Personnel Matrix
    m) Procedures for Completion Inspection
2) A chart showing the QC organizational structure and its relationship to the production side of the organization
3) Names and qualifications, in résumé format, for each person in the QC organization
4) Duties, responsibilities, and authorities of each person in the QC organization
5) A listing of outside organizations such as architectural and consulting engineering firms that will be employed by the General Contractor and a description of the services these firms will provide

6) A letter signed by an officer of the firm appointing the QC Manager and stating that he/she is responsible for managing and implementing the QC program as described in this contract – include in this letter the QC Manager's authority to direct the removal and replacement of non-conforming work

7) Procedures for reviewing, approving, and managing submittals – provide the names of the persons in the QC organization authorized to review and certify submittals prior to approval

8) Testing laboratory information required by the paragraphs entitled "Accredited Laboratories" or "Testing Laboratory Requirements", as applicable

9) A Testing Plan and Log that includes the tests required, referenced by the specification paragraph number requiring the test, the frequency, and the person responsible for each test

10) Procedures to identify, record, track, and complete rework items

11) Documentation procedures, including proposed report formats

12) A list of the definable features of work: A definable feature of work is a task which is separate and distinct from other tasks and requires separate control requirements. As a minimum, if approved by the Contracting Officer, consider each section of the specifications as a definable feature of work. However, at times, there may be more than one definable feature of work in each section of the specifications.

13) A personnel matrix showing, for each section of the specification, who will perform and document the three phases of control, and who will perform and document the testing

14) Procedures for Identifying and Documenting the Completion Inspection: This process will include in these procedures the responsible party for punch out inspection, prefinal inspection, and final acceptance inspection.

LEVEL II BAR CHART

Now that we have covered critical submittals with respect to the schedule, we can turn our attention to Level II bar charts as a tool for organizing and visualizing project activities. A simple way to think about a Level II would be in terms of each project's individual name. During the construction timeline, multiple entities must commit their resources to maintain the project schedule effectively. Initial Level II schedules help streamline planning between contractors and prevent any single party from becoming overwhelmed during construction phases. Early coordination prevents the need for significant adjustments later in the process. After establishing the sequence and estimated duration of each primary activity, we can develop a Level II precedence diagram. Each project will have its own Level II schedule.

The Level II Bar Chart is prepared from the information on the Level I. Appendix 3-C is another Level II with the major difference being the construction activities were sorted by activity number to ease the transfer of information to the Level II. Vertical lines have been placed between periods and horizontal lines have been drawn to separate the activities. The man-day estimate has been taken off the CASS and drawn to the left of each activity number. The man-days have also been written over each line representing the activity duration. We are going to now transfer the information to a Level II Bar Chart where we list the master activities in a column on the left and the periods of the entire project duration across the top. Next to each master activity we show the man-day estimate for that master activity. The next column is the weighted percent, which is the master activity man-day estimate divided by the total project man-day estimate expressed as a percent (multiplied by 100).

Once we have all bars signifying master activity durations and the man-days scheduled on the bar chart, we total the man-days scheduled for each week period at the bottom of each column. The cumulative man-days scheduled is equal to the man-days scheduled for each week period added to all previous man-days scheduled. The percent complete completion percentage scheduled is equal to the cumulative man-days scheduled divided by the total project man-days. The scheduled progress curve is then drawn by plotting the percent complete completion percentage scheduled at the end of each period plotted against the scale on the right of the bar chart.

LEVEL III BAR CHART

Once we have determined our construction schedule on the precedence network, we transfer the information to a bar chart (Gantt Chart Schedule). We do this because the scheduled dates are much easier to read on a bar chart. The project schedule is referred to as a Level III. As shown in Appendix 3-C, all the construction activities are listed down the left hand side and the time scale is shown across the top of the page. The time scale will go from the first workday of the project to the last workday, in increments of a day or week, depending on your needs. The start date, finish date, and duration of each construction activity will then be shown on the left side of the bar chart. The horizontal bar representing the construction activity durations corresponding with the time scale appears on the top. Free float is shown via blocks listed on the left side. For activities with no free float, the float will be listed as zero.

LOGIC NETWORK

The construction supervisor will develop a sequential diagram illustrating the progression of building tasks from start to finish and their interconnected relationships. Creating this logical framework during the breakdown of project activities is crucial to ensure complete coverage of all work elements. We don't have the duration of construction activities yet so we are only concerned with the sequence of work. Each construction activity is represented by an activity block. In the network shown below, activity 310 can not start until activity 100 is complete and activity 400 can not start until activities 300 and 310 are complete.

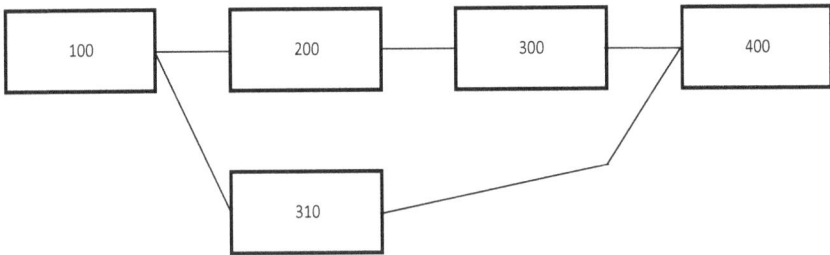

DIFFERENT LOGIC TYPES

All examples shown so far have used "finish to start logic," where an activity has to finish so the next one can start. There are two other types of logic relationships which are frequently encountered: start-to-start (S/S), where the start of the second activity is dependent on the start of the first activity; and finish-to-finish (F/F), where the finish of the second activity is dependent on the finish of the first activity. Finish-to-Start logic will give you the longest total project duration and is the most common logic type used in the construction industry. The S/S and F/F logic can be used to compress (shorten) the schedule. This compression is often used in the execution phase of the project to catch up when the project falls behind. There is also a fourth relationship, start-to-finish (S/F), but it is rarely used in construction scheduling, and we do not recommend applying it.

Note: Equations marked with an asterisk (*) are changed with different types of logic (S/S or F/F).

Start-to-Start

Forward Pass	Early start + Lag = Early start of next activity
Backward Pass	Late start - Lag = Late start of preceding activity
Free Float	Early start of next activity - Lag - Early start

Finish-to-Finish

Forward Pass	Early finish + Lag = Early finish of next activity
Backward Pass	Late finist - Lag = Late finish of preceding activity
Free Float	Early finish of next activity - Lag - Early start

The general rule to follow with different types of logic is to always follow your logic connectors.

THE BASIC SCHEDULE (FORWARD & BACKWARD PASS)

Having used contractor input to determine construction activity durations, we can now go back to our logic diagram, insert the durations, and determine our basic schedule. We will work through an example here. A few small adjustments might be needed to our initial timeline (check resource balancing) before finalizing our schedule.

On our precedence network, we will insert the activity number, description, and duration for each activity into an activity block. A typical block is shown below:

Typical Activity Block		
Activity Number	Activity Duration	
Early Start	Activity Description	Early Finish
Late Start	Activity Resources	Late Finish
	Total Float	Free Float	

To begin creating our basic schedule, we start with a forward pass. We input zero as the early start date for the very first task. Next, we add the task's duration to its early start to calculate the early finish. This early finish becomes the early start for the following task. It's important to note that for activity 400, which has two preceding tasks, we select the larger of the early finish dates as our critical path activity.

Early Start + Duration = Early Finish
Early Finish + Lag = Early Start of Next Activity

As you can see looking at the network on the network, the early start and finish dates for any activity depend on the number and duration of the activities which have to be done before it. Next, we are going to do a backward pass. We start by taking the early finish date for the last activity and making it the late finish for the last activity, since the critical path has zero float. Subtracting the duration for each activity from the late finish date will give us the late start date. The late start date will become the late finish date for the preceding activity, this time subtracting any lag. Here is the equation for figuring late start and finish:

Late Start - Duration = Late Finish
Late Start - Lag = Late Finish
of Preceding Activity

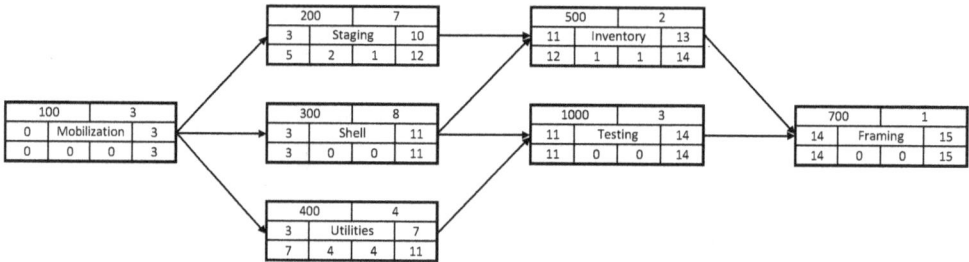

200		7
3	Staging	10
5	2	1	12

500		2
11	Inventory	13
12	1	1	14

100		3
0	Mobilization	3
0	0	0	3

300		8
3	Shell	11
3	0	0	11

1000		3
11	Testing	14
11	0	0	14

700		1
14	Framing	15
14	0	0	15

400		4
3	Utilities	7
7	4	4	11

TOTAL FLOAT

Total Float, also referred to as Total Slack in Microsoft Project, represents the number of days an activity can be delayed without affecting the project's end date. Examining activity 200 in the diagram, we notice it could conclude as soon as day 10 or as late as day 12, indicating a two-day total float for this activity. The two-day flexibility we have in finishing activity 200 anytime between day 10 and day 12 is termed Total Float. We determine total float by deducting the early finish date from the late finish date (or the early start date from the late start date).

Total Float = Late Start – Early Start
(or Late Finish – Early Finish)

FREE FLOAT

Free float represents the number of days an activity can be postponed without impacting the next task's timeline. In other words, it's the delay an activity can have without pushing back the early start date of the following task. To calculate free float, we take the early start of the subsequent activity, subtract any lag, and then subtract the early finish of the activity we're assessing. This method helps determine how much flexibility exists within the schedule for each task. It is important to understand that there is a difference between free float and total float.

An important distinction here is between lead time and lag. Lead time is essentially the amount of time you can accelerate a successor activity. So, if task 1 has a 10-day duration and its successor activity (task 2) has a lead time of 3 days, task 2 can start 3 days before task 1 finishes. Lag time also refers to a delay based on the preceding activity, but it's the amount of time the successor activity has to wait to start after its predecessor is complete. That would mean that task 2 in the prior example could not start until day 13, or 3 days after task 1 is complete.

To calculate the free float for activity 200 above, we would take the early start for activity 500, subtract any lag between 200 and 500 (zero in this case) and subtract the early finish for activity 200 (11-0-10=1). Activity 200 has a free float of 1 day. It's evident that a one-day delay in activity 200 won't affect activity 500's early start date. If activity 200 is delayed by 2 days, activity 500 will need to start on day 12, reducing its float by 1 day (reaching zero in this instance). Any delay beyond 2 days for activity 200 will impact the project's completion date since activity 200 only has 2 days of total float.

Free Float = Early Start (Next activity) – Lag (if any) – Early Finish

CRITICAL PATH

Examining activity 200 above, we notice it could begin as soon as day 3 or as late as day 5. We've calculated the total float as 2 days by subtracting 3 from 5. When early and late starts match, there's no float; we must start that activity on its early start date. Delaying it would delay the entire project. Activities without float are considered critical. The first and last activities are always critical, with a critical path of activities connecting them. In the earlier example, the critical path is 100-300-1000-700. This path helps management focus on activities that can't be delayed and allows other tasks to float until reaching their free float limit. The critical path enables managers to prioritize tasks that directly impact project completion.

RESOURCE LEVELING

Resource leveling involves aligning planned construction tasks with the available workforce. The aim is to keep the crew productively engaged daily while ensuring they can meet the scheduled workload without delays. To achieve this balance, we need a set crew size, a time-based schedule, and a graph showing the daily manpower requirements for scheduled tasks. The first page of the Level III resource leveling chart can be found in Appendix 3-B.

Looking at Appendix 3-B, we see the resource histogram at the bottom of the page, which is the original number of personnel required to complete the critical activities scheduled for each day.

The primary task of resource leveling is to schedule the non-critical work as you have people available to do the work. The crew sizes for each non-critical activity have also been penciled in next to the activity start date. The total crew size in this example is seven. We have resource leveled this project for a small crew scenario. We began the resource leveling process by committing to doing the critical path as shown and plugging the resources into each activity. The critical path will obviously not keep the entire crew busy in this case. The non-critical activities shown on their early start dates, but we may need to delay them if we don't have people to start them there.

When postponing a low-priority task, we aim to begin it immediately once we have staff ready to tackle it.

Appendix 3-B has not been resource leveled. The activities were scheduled beginning with activities with the least amount of total float to those with the most amount of total float. This process would be continued through the rest of the project. Note that we did not have the personnel to start all the activities at once. Some minor adjustments on crew sizes and durations may be required to ensure full utilization of assigned crew. Once all the activities are scheduled, then we can input the non-critical resources and delayed start dates (using lag) and create a new bar chart.

COMMON REPORTING

1) Banded Schedule: This is your most common view of a schedule. It will contain all activities, sorted by start date, and aligned under their respective work breakdown structure (WBS), which are headers used to encapsulate a series of activities in your schedule.

2) Critical Path Schedule: This is a version of the schedule that only shows the activities on the critical path to completion. Critical path activities cannot be delayed without negatively affecting the completion date of the overall schedule.

3) Narrative Report: This takes what the schedule is showing and puts it into narrative format. It would identify written critical path activities, how much the schedule has gained/slipped, logic changes, duration changes, etc. An example of what is contained in the narrative report can be found in Appendix 4-6.

4) S-Curve: A tool to visually identify whether a project's expenses are front loaded or not. The Owner is not going to want to spend all the project cash at the beginning of the project and lose leverage. A proper S-Curve will have a true S shape where one axis is cost and the other is time, showing a gradual increase in expenses before leveling off.

5) Cash Flow Analysis: This shows approximately how much cash you can expect to receive for each period if the project is built according to the schedule.

6) Longest Path Schedule: Simply put, this schedule shows the activities that are the longest path to completion. It is often confused with the Critical Path schedule given that at the outset of a project the two have the same expected completion date. But as delays in a schedule occur, it is possible for multiple Critical Path schedules to appear. The key differentiator, though, is that not all of these changes affect the Longest Path schedule. Whichever critical path schedule leads to the latest anticipated completion is what we correlate to the longest path schedule.

7) Three-Week Lookahead: This schedule shows activities that have started, or are projected to start, in a three-week timeframe from a prescribed start date.

8) Weekly Production Report: This report is mainly used only when your schedule is loaded with a manpower resource to empower the project leadership to make decisions that can affect the project's profitability. The labor report is a key building block in the success of a project. Your willingness to use it, and use it right, will determine your success or failure.

SOFTWARE

In the world of construction management, there are many tools at our disposal to help track and manage the project schedule. This is important because next to material, time is one of the largest expenses you can incur. Every day beyond the expected completion date leads to additional labor, insurance, and other overhead related costs, not to mention potential fines and penalties for failing to meet your contractual obligation. Every project will vary on what management system will work best for based on the overall complexity and comfort level of the project managers, so there is no right or wrong answer and our intent in this chapter is to remain neutral.

Microsoft Excel

This is an application that has been around forever and one that nearly everyone in the professional world either has on their computer or has at least used at one point or another. As with all the options we will look at in this chapter, there are both pros and cons to its role in project management. On the positive side, it is fairly well-known, easy to use, and seamlessly integrated with the rest of the Microsoft Office suite. The ability to program formulas into individual cells is also a tremendous advantage when performing the Manual Earned Value calculation. And because it is so base level and generic, Excel can be adapted for a wide range of projects with just a high level understanding of the formatting features. On top of all that, it is free, or at least included in the price of the office suite you likely already have.

On the negative side, Excel can become much more time consuming than a more sophisticated offering. There are no built-in Gantt charts, so all visual bar chart representations must be manually created. A hidden benefit of this extra work, though, is that multiple fields can be created for any data type and fields can easily be added or modified. Manual Baseline will need to be calculated either by drawing a line or using an Excel function instead of automatically populating like it would in other programs. Excel does not allow for resource management because it will always assume an unlimited number of resources are available and all activities must be managed manually due to the non-dynamic nature of the program. The last drawback is that the master file is stored on an individual computer with no central server based sharing, although through Microsoft 365 this has improved.

Microsoft Project

As another Microsoft product, Project offers seamless integrations with the other suite of products. You have the ability to cut, paste, or otherwise import and export data as you see fit, which can save a lot of time from other methods. With direct export to Excel for the majority of data, the output can easily be formatted into graphical display options such as tables, bars, profiles, and spreadsheets without needing the level of programming required if starting off in Excel. Additional Microsoft features that can make Project a smart choice for savvy users surround ease of use for anyone new to scheduling. Wizards are step-by-step tools used to help train and guide new users, combined with the Microsoft foundation that now integrates with other software like Enterprise Project and eLabor as well.

The benefits of Project continue with the customization options that allow for data presentation in almost any manner, with any information, thanks to the multiple fields available for almost any data entry. Unlike Excel, Project is capable of Earned Value calculations with automatic visual presentations (black line through bar) and quantifiability (WIP %). Because of these flexible data entry fields you are also able to establish multiple baselines and update them as needed to reflect the dates you want to freeze panned activity characteristics. Another feature Project has that Excel lacks is the capability of managing almost any resource. Even though Project assumes unlimited resources, it also allows leveling to prevent resource overallocation. An example for this functionality is the ability to support multiple billing rates. This is especially useful in the analysis phase for the manager to make a cost-benefit proposal by calculating the total cost of the project using the different rates for the different resources.

And the benefits don't stop there. Project is also server-based in addition to having stand-alone capabilities, which helps to reduce the cost by limiting licenses. Although with a total cost of $400 to $500 for a full license, the thought of paying for additional seats is not the worst thing in the world. For that price tag, Project proves to be a tremendous value with its ability to manage projects with activities ranging from 2 to 2,000 activities. All this with the added bonus of organizing schedules via the use of precedence links, rather than dates, so the schedule can update automatically when changes occur.

With such a robust set of benefits it might seem like Project is an ideal solution, but there are also several drawbacks to take into consideration. While data entry is user friendly, it can also lead to mistakes because it is very easy to accidentally add verbiage, change durations, logic, etc. It can also be difficult to produce necessary reports from Project without creating custom reports. The "default" options in Project can lead to inaccurate schedules due to the differences between manually scheduled and auto-scheduled activities. Creating unique ID numbers is cumbersome, but not impossible. And last, and possibly least depending on how visual you are, Project is not as easy on the eyes as other alternatives. In order to get the bold colors, distinct headers, and other features that come standard with other software, you will need to spend some time on the configuration.

Primavera P6

Primavera P6 is another program that is fully compatible with the Microsoft Suite, although beginners should expect a moderate learning curve due to its complexity. Unlike some general-purpose scheduling tools, P6 is widely used across many industries, including engineering, manufacturing, aerospace, and defense. However, it is particularly powerful in the context of construction management, where its features provide strong benefits for Project Managers. Like Microsoft Project, P6 supports Earned Value calculations and the ability to set multiple baselines. Its resource management capabilities are well suited for construction applications, requiring less manipulation when compared to software that assumes unlimited resources. P6 also provides dynamic scheduling capabilities through the use of precedence links, making it an effective tool for managing complex construction projects.

Because this is not a proprietary Microsoft product, it is not as widely used or accepted, meaning it requires a keen understanding of scheduling principles for someone to be able to comfortably use it. What could be seen as a benefit with this program in its robust applications also tend to be a disadvantage for smaller companies, or companies that only work with smaller projects. As a construction specific program there is a much smaller pool of potential users than the other options, which means the Company needs to recoup their research and development costs across a smaller pool of users, increasing the cost of a license to anywhere between $2,500 to $8,000. The ability to operate on either a server or individual computer does allow for the reduction in additional license costs though.

NETWORK CALCULATIONS

While most scheduling software has the necessary algorithms built-in to perform these calculations simultaneously, it is still important for any manager or superintendent to have an understanding of the math and logic being used behind the scenes. A calculator can reduce the time needed for mathematical calculations, but a basic understanding helps the user know when the output generated does not look right and can prompt them to inspect the inputs, the same holds true for scheduling calculations.

Forward Pass

ES + Duration = EF

EF + Lag = ES Next

Note ES_D: $EF_C + LAG_{CD} > EF_B + LAG_{BD}$

Backward Pass

LF - Duration = LS

LS - Lag = LF Previous

Note LF_C: $LS_E - LAG_{CE} < LS_D - LAG_{CD}$

Total Float

LS – ES or LF-EF

ES Next - LAG - EF

Note FF_B: $ES_F - LAG_{BF} < ES_D - LAG_{BD}$

Logic Relationships

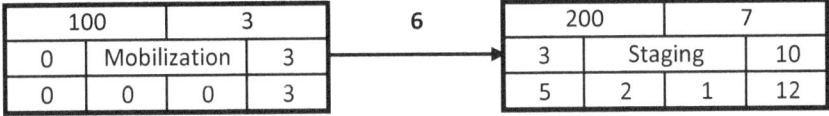

100		3
0	Mobilization	3
0	0	0	3

→ **6** →

200		7
3	Staging	10
5	2	1	12

Start to Start

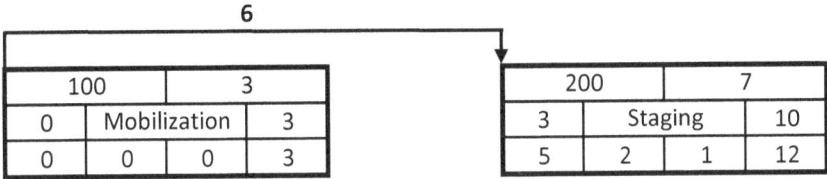

6

100		3
0	Mobilization	3
0	0	0	3

200		7
3	Staging	10
5	2	1	12

Finish to Finish

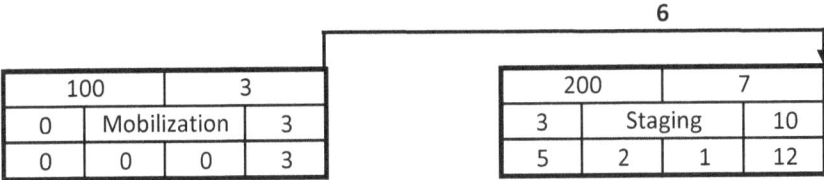

6

100		3
0	Mobilization	3
0	0	0	3

200		7
3	Staging	10
5	2	1	12

1	2	3	4	5	6	7	8	9	10
Place Exterior Studs			Plumbing						
4	4	4	2	2	2				
Place Exterior Sheathing			Heating			Insulation			
4	4	4	2	2		3	3	3	3
Place Interior Studs			Electrical						
	2	2	2	2					
8	10	10	6	6	2	3	3	3	3
Total Number of Workers Per Day									

CHAPTER 5

JOBSITE MANAGEMENT

In this chapter we will discuss ways to help organize your construction site for maximum efficiency. This includes materials, tools, job site appearance, Customers, field offices, and initial set-up. It should come across to most as common sense, but a clean and safe jobsite will make every aspect of your project go smoother. Cleaning up loose debris like nails and metal shavings will reduce injuries. Organizing stacks of material in their appropriate places will make it easier to perform inspections on the quality of workmanship. Fewer injuries and better inspections will ultimately lead to timely completion schedules.

Every project is built around four essential functions: safety, quality, schedule, and budget. The order of importance among these functions is critical. Safety comes first — if someone is injured or killed, no one will care about the quality of the work or how quickly it was completed. Quality follows closely behind, because years after a project is finished, stakeholders will judge it by how well it performs, not by how fast it was built. When safety and quality are effectively managed, the project team is better positioned to meet schedule and budget objectives. In this way, prioritizing safety and quality ultimately drives productivity and reduces the risk of overruns. See Appendix 5 for a sample checklist on specific site layout objectives, and Appendix 14 for a Preconstruction Checklist to guide your planning before construction commences.

DOCUMENTATION

Immediately at the start of the project, one complete set of plans, specifications, and addenda is to be sent to the job site clearly marked as "As-Builts." There is no need to post the addenda; just include it as part of the set. These documents are *not* to be used for construction. They are to be properly filed and kept in good condition. Review the contract documents to determine any specific conditions required by the Owner for preparation, maintenance, and delivery of the As-Built Drawings. Comply in every respect. It is the responsibility of the Project Engineer, Site Superintendent, and other field staff to verify that any deviation between actual construction and the original design is, in fact, properly authorized and documented. These will include items such as:

1) Approved change orders
2) Clarifications not involving cost or time
3) Accommodations of field conditions that are slightly different than those originally anticipated
4) Actual locations and configurations of existing underground lines and construction as they are uncovered during the course of the work
5) Actual locations of new underground work if at all different from the plan locations

6) All dimensional deviations and all references to the detailed change records on *both* the job-site document set and the As-Built Drawings *as they occur* in sufficient detail. All such additions, deletions, or changes are to:
   a) Be indicated in red pencil or pen
   b) Be dated
   c) Include *clear* reference to appropriate authority for the modification, such as:
      (i) Change order file number
      (ii) Job meeting item number
      (iii) Conversation and memo of confirmation with name, conditions, etc.
      (iv) Structural Modification Authorization Forms

In the case of change orders and detailed clarifications, it is not necessary to redraft the details of the change. Cloud the area affected by the respective change or clarification, and reference the appropriate change order number or other complete reference. Whenever possible, tape a photocopy of any available sketches on the contract set. Most likely, the only available space will be on the back of the previous page. In this case, simply note the location on the modified plan. Include copies of any Structural Modification Authorization Forms. It is the Project Manager's and the Superintendent's responsibility to police each major Contractor or trade Contractor to include their own as-built information on the Company field set and As-Built Drawings *on a weekly basis*. The Project Manager, as an express condition of payment, should confirm this information monthly. These Contractors include at a minimum:

- Concrete
- Structural steel

- Plumbing
- HVAC
- Fire protection
- Electrical
- Controls
- Communications

Confirm final as-built configurations required by the contract prior to delivery to the Owner. It may, for example, be required to transfer the information to a set of plans provided by the Owner. In such cases, it is not necessary to transfer any supplemental documents taped to the plans. The references will be adequate. Include all engineered layouts, confirmations, and certifications provided and all certified As-Built Drawings by all trades required to provide them (fire protection, for example). Hand deliver the completed as-built documents to the Owner, and have them signed for by an individual authorized to receive them.

PRECONSTRUCTION CONFERENCE MEETING AGENDA

Date: 06/25/25

Project: Fort Valor Command and Control Center

Contract No.: RCP-2025-001

Attendees: See Attached List

Introduction

The purpose of this conference is to make sure that everyone involved understands the work required, project conditions and procedures in order to complete the work with a high-quality standard, on or ahead of schedule, and within budget.

1. **Contract Date**: 05/15/25
2. **Official Start Date**: 01/01/25
3. **Official Substantial Completion Date**: 12/31/27
4. **Project Personnel**:
    a. **Design/Builder**: Horizon Builders, Inc.
    b. **Project Manager**: Jessica Miller
    c. **Project Administrator**: John Davis
    d. **Contractors**: See attached list of contractors/contacts
    e. **Owner Representative**: US Army, Ft. Valor

Review of General and Special Conditions/General Contractor Requirements

Job Site

1. **Security**: The General Contractor is responsible for reporting losses to field representatives and local law enforcement officials.
2. **Storage**: The General Contractor is responsible for identifying the designated laydown yard for storing all project-related materials, equipment, and supplies.
3. **Parking**: The General Contractor designates parking areas for contractors, subcontractors, and visitors, as well as identifies locations for jobsite facilities such as trailers or offices in relation to parking and access points.
4. **Sanitation Facilities**: The General Contractor coordinates the provision of adequate sanitary facilities for all workers in compliance with OSHA and local regulations, identifies locations for portable restrooms or permanent facilities, and develops a plan for maintenance and cleaning throughout the project.
5. **Clean-Up**: The General Contractor ensures trash and debris are removed daily. Keeps trash removed from the exterior and picked up (subcontractor will do their own clean up).
6. **Inspections**: The General Contractor is responsible for obtaining all inspections required by the specifications, testing agencies, and/or Client agencies.
7. **Safety**: Each contractor is responsible for taking proper safety measures. Refer to AIA A201, Article 10.
   o **Special Safety Requirements:** Floor scanning is required prior to drilling into existing floors due to in-slab electric.

8. **As-Builts:** The General Contractor will maintain one clean set of drawings marked "As-Builts" on which all changes are indicated immediately as the work progresses. All changes and clarifications should be attached to this set as soon as issued.

9. **Project Meetings:** Monthly Progress Meetings will be held on the 4th Wednesday of each month at 2:00 pm.

Submittals Required from General Contractor

1. **Shop Drawings and Samples:** Refer to the submittal section of your construction specifications.

2. All submittals are to be delivered to General Contractor through the Contractor.

3. All shop drawings/submittals must be initialized by the General Contractor to indicate that he has reviewed the material. Shop drawings/submittals received without General Contractor's review will be returned without action.

4. Forward submittals to the General Contractor, attention: [Project Administrator]

5. Include a transmittal (or letter) with each submittal listing items submitted, and designating specification section covering each item submitted, e.g.,
   o Trade-specific certifications (e.g., welding)
   o Trade-specific shop drawings (e.g., roofing and sheet metal)

6. Include the following:
   o Project Name
   o General Contractor
   o Project number

7. The General Contractor will keep three sets of each submittal and return the rest. Allow a minimum of fourteen calendar days for submittal processing.
8. Material color selection/submittals for all interior finish colors shall be coordinated by the General Contractor to be submitted in one interior finish color selection package.

Administrative

1. **Correspondence**: Address all correspondence to the Project Manager and submittals, applications for payment, change order proposals, etc., to the Project Administrator as follows:
   a. Attn: [Project Manager/Project Administrator]
   b. Project: Ft. Valor Command and Control Center
   c. Project No.: RCP-2025-001

2. **Cost Breakdown and Progress Schedule**
   Cost Breakdown: The General Contractor is to break all work items into tasks outlined on the schedule (including proportionate share of overhead and profit for each line item.) Submit cost breakdown within 5 days of execution of the contract. Once approved, this cost breakdown will remain unchanged (except for change order items which will be listed on the last page) until contract completion.

3. Applications for Payment

Payment applications must be sent to the General Contractor:

a. Use AIA Forms:

    a. G702 – Application and Certificate for Payment (has a signature line for the design/builder)

    b. G703 – Continuation Sheet

b. Break payment request into division incorporated, and materials stored (if applicable). If off-site storage is permitted, proper documentation (invoices, insurance) must accompany the request. Make sure the pay request is properly signed.

c. Payments cannot be made on change orders until all parties formally approve them.

d. Submit electronic copies of pay applications to the General Contractor; be sure all figures are legible, and those figures have been properly checked. Simple math errors may be corrected as the pay requests are processed.

e. Final payment requests will not be processed until work is complete, including submission of all required closeout documents.

f. Each pay application must be received by the General Contractor no later than the 25th of each month. Late submissions will be held until the following month. Payment will be made by the 25th of the preceding month.

PERSONNEL

Types of personnel needed for jobsite management can vary from one project to the next depending on size, scope, location, and numerous other external factors. Obviously the same holds true for the specialty of the tradesmen needed, but there will be some common roles almost always required:

- **Project Manager:** They are responsible for more of the administrative side of supervision. Most of their time is spent dealing with subcontractors, paying invoices, communicating with the Owner or their representative, maintaining the schedule, and tracking change orders and approvals.
- **Superintendent:** This person is responsible for the physical day-to-day operations, safety, environmental issues, and overall quality assurance. They are the drivers of the schedule and make sure what is being built matches what is shown on the drawings.
- **Quality Control Manager:** They assist the superintendent and project manager by focusing solely on the quality control aspects, ensuring submittals and documentation match the materials on-site, materials are installed according to specification, as well as testing and commissioning materials and equipment.
- **Safety Officer:** Responsible for assisting the superintendent and project manager with overall safety. They will inspect all equipment to make sure it is in good working order and observe jobsite conditions and behaviors for adherence to the outlined safety policy.
- **Environmental Officer:** They oversee all hazardous material, hazardous waste, and erosion control measures.

- **Owner's Representative:** The liaison between Owner and General Contractor, generally an independent third party serving as a check and balance measure to make sure the end user is getting a product that matches what has been designed and paid for. They are usually a specialist in the type of building being constructed so they understand the subtle nuances that may make a material difference in the finished project.

MATERIAL

There are a lot of things to consider when setting up the construction site. Material is one of the most important because there will be a large quantity to manage, which means storage must be planned. Depending on the project, some questions you need to ask yourself before mobilizing material or scheduling deliveries are:

- How big an area do you have to store material on-site?
- Can the material be secured?
- Is the material exposed to the weather?
- Do you have too much material on-site?
- Is the material stored properly?
- Do you have SDS on the material that requires it?
- Is it the right material?
- Do you have to build a material storage area and make it in an area that is not going to be in the way of construction?

TOOLS

Tool accountability is one item you as a project manager need to stay on top of. The simplest way to ensure you have all the tools you need is by conducting a thorough inventory assessment at the beginning of all jobs. Periodic audits of tools will ensure tools don't "walk off."

JOBSITE APPEARANCE

The appearance of your jobsite, well put together or disorganized, is the first thing anyone is going to see. You can be doing high quality work, but if your jobsite looks disorganized that is the only thing people are going to remember. Making sure the crew members are picking up after themselves during the workday and having a final cleanup at the end of each day is key to a successful project. Simply put, a clean job site is a safe job site.

CUSTOMERS

One of the most important impressions made upon Customers is the one made by the Project Manager when presenting their job. Know all of the ins and outs of the project so you can speak positively and intelligently. Be professional — first impressions mean a lot. Describe the project in general, including the type of construction and finishes. Mention specific safety measures taken and any special quality control concerns. Explain the project schedule using your knowledge of the schedule and schedule reports. You want the Customer to leave with an impression that you know what you are doing and have everything under control. Do not try to bluff your way through anything. If you do not know, then state so and get back with the Customer with the correct answer. It is understandable not to know everything off the top of your head, and your honesty will be appreciated. Remember, an impression, whether good or bad, goes with every person with whom you come in contact.

FIELD OFFICE

The field office may be an equipment shelter or an appropriate structure to provide the necessities for a product work environment: heating and cooling, shelter from the elements, a meeting space, a working space, and if you are really fancy – a restroom. A field office should have the folowing items on display:

1) Level II Bar Chart
2) Safety Plan
3) Quality Control Plan (QCP)
4) Construction Activity Summary Sheets (CASS)
5) Three-Week Look Ahead
6) Emergency Phone Numbers
7) Project Folder
8) Clean set of Working Drawings (Red Lines drawn in)
9) Project Daily/Weekly Reports

Jobsite Information Board

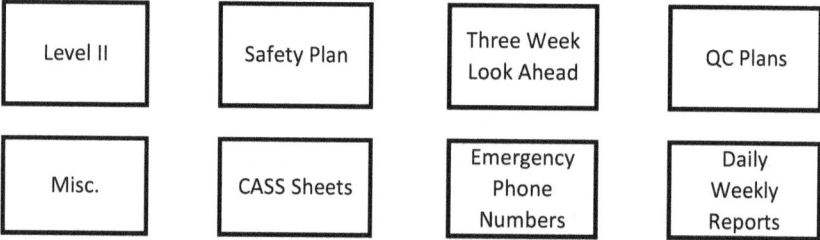

Level II	Safety Plan	Three Week Look Ahead	QC Plans
Misc.	CASS Sheets	Emergency Phone Numbers	Daily Weekly Reports

INITIAL SET-UP

There are many things to consider when setting up your jobsite. It is best to give this a lot of thought before ever breaking ground or placing the first order because you will have a blank canvas to work with. Once the job site is crowded with people, vehicles, equipment, and material, it becomes harder to navigate changes, which can lead to costly delays. Here is a list of the top considerations we always advise paying attention to:

- **Access:** Can all the trucks we need to use get into the site? If not, do we need to build them access roads? Can employees in their personal vehicles gain access to the site or do they require a special mode of transportation? Ensuring emergency vehicles can have unimpeded access is also important.
- **Haul roads:** These access routes are necessary for moving materials from one side of the site to the other.
- **Office trailers:** Depending on the size of the jobsite, these could be in a centrally located area, at the entrance to the jobsite, or near the entrance to the building.
- **Restrooms:** They can be located in the individual trailers or near them when possible. On larger sites you will want to have them strategically located throughout.
- **Potable water:** You need to determine where both the temporary and permanent connection points will be.
- **Electricity:** Same premise as the water connections.
- **Material management:** We refer to the area where materials are stored as the lay down yard and you will want to make sure it is accessible to all areas of the project or

designate several locations where various working supplies of material can be allocated.

- **Parking for equipment:** This aligns with material management in making sure everything is secure and easily accessible.
- **Signage:** There are too many types of signs to list, but most job sites include ones like visitors must sign in, danger overhead electrical, speed limit, hard hat area, etc.

INSPECTIONS

The biggest inspection a Project Manager can expect is from the End User (Customer), and to be prepared for that, they should already be inspecting their jobsite on a daily basis for basic items that will keep the site running efficiently and safely. The checklist in Appendix 4-1 will assist the project manager in preparing for an inspection, but there are four main areas to focus on:

- Safety Items (both physical and behavioral)
- Housekeeping
- Crewmembers
- Project Management

CHAPTER 6

MATERIAL MANAGEMENT

Material management and accountability is the responsibility of the subcontractors and self-perform crews, but the Project Manager is responsible to ensure the Subcontractors manage their material resources appropriately. You have already learned how material is tied to productivity, now in this chapter you will learn how to identify long lead items and track materials throughout the project schedule.

FIELD STAFF

We already covered many of the different roles you will need to fill back in the last chapter on jobsite management, but now we are going to focus more specifically on how those roles fit into the material management process:

- **Superintendent:** Coordinates deliveries, ensures locations are clear to receive shipments and there are personnel assigned to unload, and collects any shipping paperwork.

- **Safety Officer:** Oversees the stocking of first aid kits, fire extinguishers, signage, and any other job specific safety equipment needed. The Safety Officer also ensures that materials are delivered and offloaded/handled safely, and that the laydown yard remains safe and free from hazards.
- **Environmental:** Procures straw bales, silt fencing, aggregate to slow down water runoff, temporary seeding, erosion control mats, and any other job specific environmental equipment needed.
- **Quality Control Manager:** Ensures what has been brought to the site has been approved in a submittal or shop drawing, verifies in production testing, inspections of welded steel, and approval of any other specific parts or equipment brought on the site.
- **Owner's Representative:** With material management, they act very much in the same way the QC Manager does, only through the lens of the Owner's best interest.

PRECONSTRUCTION RESPONSIBILITIES

This is your time to plan for your project and identify the resources you need, before construction has begun. It is your responsibility to ensure you know what materials are needed and when they are needed.

Bill of Materials

After plans and drawings are drawn up and approved, the Bill of Materials (BOM) is created. The BOM must include everything you need to complete the construction. This may include everything from screws to steel beams and all the tools and equipment needed to then install the materials. This material must be accounted for before a start date can be scheduled or you can find yourself wasting valuable man hours as they stand around with nothing to work on. Your number one job prior to construction beginning is to ensure the BOM is correct, so you can bid and award to the appropriate sub, resulting in minimized change orders due to unidentified scope.

There are three commonly misunderstood terms when it comes to the BOM:

- Rework (Appendix 6-1): Used for correcting a deficiency on-site.
- Warranty (Appendix 6-2): Correcting deficiencies after the crew has left site.
- Add-ons (Appendix 6-3): A post shipment not resulting from an omission or deficiencies. These are strictly unexpected material requirements bred from a necessity that is not caused by the failure of Company personnel.

Prior to addressing a warranty, add-on, or rework, a detailed explanation must be created via the specific form for the issue at hand. These must be approved in order to fund the add-on, rework, or warranty. Examples of these forms are provided in the Appendix.

Tracking Materials

Once you have confirmed your subcontractor has ordered your materials, you can stay on top of tracking by simply checking in with them via weekly foreman's meetings. It always helps to stop by or call your respective subcontractor and confirm or track your material shipments between weekly meetings if they are waiting on items that are crucial to advancing the project schedule or negatively impacting other subcontractors. This not only makes sure your shipment arrives on time, but it also gives you face time with the people who will be doing the work.

Money Management

A key area of your control is money management. When a project is sold, the estimating department creates an estimate that your budget is based on. Each piece and part whittle away at the budget, so you need to be concerned about waste, material inflation, scope creep, buy-out, etc. Every hour or day spent on tasks you had not allocated for in the bid quickly adds up to wasted money as well. Once you have exhausted your budget, you start reducing profit margins for your own Company and the Client that hired you. As soon as the profits of the project begin declining, the individual stakeholders are going to assume the worst. No one wants to be part of an insolvent project where they run the risk of not getting paid.

On-Site Construction

Planning is done and you are now ready to start your project. Your success is dependent on how well you have chosen your Subcontractor base and the conciseness of your scope bought into each buy-out. Scope inclusion is a fine science that takes experience and a keen sense of what may occur in the future. Fully understanding all your plans is the first step in ensuring the scope is as inclusive as possible. It is also important to do constructability reviews to ensure future amendments to the plans do not affect your budget in a drastic manner.

Storing of Material

If you followed the best practices from Chapter 5, when we discussed material storage as part of job site management, then this should just be a reminder to store materials as close to their final installation locations as possible. Keep in mind the possibility of theft and the security requirements of the material. On many jobs the material is required to be "specially controlled" to prevent tampering with materials. Damage of material should be considered as important of a concern as theft since unusable material is about as valuable as missing material. Material stored should be in a place to prevent water damage, insect infestation, and damage from movement. Storing is a critical factor that should be addressed with each Customer. Proper storage facilitates efficient installation.

KEY CONSIDERATIONS

Excess material above the attic stock (the extra parts outlined in the "specifications" section of the project scope, which is created to account for parts breaking, going missing, etc.) need to be offered to the Customer/owner before they are disposed of or redeployed to another jobsite. These items have been paid for by the Customer and belong to them. Excess materials need to be noted and signed for on the project close-out report by the Customer. Even your basic stationery and office supplies like pens, notepads, and printer ink need to be identified and signed over to the Customer.

Hazardous Material

Ensure any hazardous materials you use on your project have a SDS readily accessible. All guidelines for use and storage must be followed scrupulously.

Shelf-Life

Ensure all items with a shelf life are identified and tracked to prevent possible waste due to expired shelf life.

LONG LEAD ITEMS

You probably already have a decent idea of what long lead items will be required for your project from the initial review of the drawings and completion schedule, but this becomes more critical here while you are laying out the storage and staging areas. If you checked how long it was going to take to get some of the more time-constrained items earlier in the process, check again now to ensure those times have not changed. The logic and duration for each should be properly conveyed in the project schedule for the purpose of meeting drop dead dates by having these items on-site in a timely fashion without clogging up the lay down area with material not needed in the foreseeable future.

CHAPTER 7

EQUIPMENT MANAGEMENT

As a project manager, it's important to have a solid understanding of the proper care and maintenance of the equipment being used on your project. In this chapter, we will discuss some of these concerns and issues. At a bare minimum, every operator of equipment should ensure:

- Equipment is operated in accordance with established procedures and all safety precautions are rigidly observed.
- Movement of personnel is restricted to approved trips and legitimate work purposes.
- Every construction and material transport vehicle is designated for specific building operations. Personnel transport is prohibited with this equipment. The seating capacity of any operational construction machinery determines the maximum number of occupants allowed.
- Personnel assigned to operate automotive, construction, or material handling equipment will be qualified and licensed.
- Equipment is made available for preventive maintenance service as scheduled by the operating procedures.

- Personnel operating automotive, construction, or material handling equipment perform operator maintenance.
- Personnel are familiar with current Company policies for use of equipment.
- The equipment is not used to store tools, materials, or personal gear.

FIRST ECHELON MAINTENANCE

It is the operator's responsibility to ensure proper maintenance is performed. Every operator must keep the assigned vehicle or equipment clean, safe, in serviceable condition, and perform daily operator maintenance. Daily inspections must record any defects, which should be corrected before a serious breakdown or mishap occurs. Numerous machines require regular oiling at specific spots throughout the day and the operator must handle the lubrication process. Managers should verify maintenance follows specifications detailed in the equipment manual. Even though most equipment belongs to the subcontractor, if it is broken, it will affect our ability to get work-in-place.

The operator must be familiar with equipment and use their senses to detect possible problems that occur during operation. The sense of smell (burning), hearing (unusual noise), sight (instruments), and feeling (drag, pull, vibration) are all used to detect potential problems. The operator should be ready at any time to shut down equipment if they have indication of failure. After operation, the operator should perform the established shut down procedures as prescribed in the operators' manual or other service directories. Protection of equipment must be integrated into shut down procedures to prevent theft or damage.

PREVENTIVE MAINTENANCE

Preventive maintenance describes scheduled activities aimed at optimizing equipment uptime and reducing unplanned repair expenses. Preventive maintenance encompasses safety and operational inspections, lubricating procedures, and basic servicing and calibrations beyond standard operator upkeep tasks. The standard interval between Project Manager service inspections is forty working days. It is up to the supervisor to determine if the Project Manager interval should be widened or narrowed. Even though most equipment belongs to the Sub, if it is broken it will affect our ability to get work-in-place.

SAFETY STAFF

If it has to do with safety, safety staff are the folks responsible. Some people think safety is limited to enforcing hard hat requirements and construction appropriate attire, but the scope of the safety staff runs much deeper. They don't just look out to ensure everyone working above a specific height is wearing their safety harness, they inspect and test those harnesses. The same goes for air monitors, back up alarms on large vehicles, turn signals, operational fire extinguishers, and so on. Simply having the required safety gear and equipment on-site is not enough, and it is their job to reject all defective equipment and report it to the Superintendent and Project Manager.

CHAPTER 8

PROJECT EXECUTION

Effective construction project execution demands the convergence of diverse resources at precise times and locations. This involves not just any materials, equipment, and workers, but the right ones for each task. Our job as project managers is simplified because we've already pinpointed our needs for each activity during the planning phase. We've identified the necessary tools, machinery, materials, and staff for every construction task. This chapter explores the various techniques we employ to monitor these resource requirements, starting from the initial planning stages right up to the day we begin work on a specific activity. Our goal is to ensure smooth, efficient project execution.

CONSTRUCTION ACTIVITY SUMMARY SHEETS (CASS)

Effective implementation of a CASS significantly minimizes the risk of construction delays or stoppages caused by resource shortages. While a CASS serves multiple purposes throughout different construction phases, during execution its primary role is maintaining team productivity. Most resource needs listed on a CASS typically demand additional follow-up actions from the Project Manager. These necessary actions can be monitored directly on the CASS itself. Highlight the action required, whether it is a requisition to be submitted or anything else. List the required action and the due date on the CASS and circle it in yellow. Lead times need to be considered when identifying possible problem areas.

LEVEL I, II, III BAR CHARTS

A precise and up-to-date evaluation of the project's status must be consistently maintained at the jobsite. Even a deviation from the schedule of only several days makes a big difference to a concrete supplier, a hired crane, or even a contractor. This does not mean your project has to be re-planned every week; just updating the project status can be reflected on the posted Level I, II, and III bar chart. The critical path on this bar chart should be highlighted in red and the daily status should be tracked each day. Daily status will show where you stand compared to the schedule. The Level I, II, and III bar charts demonstrate a technique for reflecting total project status on a reporting period basis. A vertical line is drawn at the end of each reporting period. The line shows at a glance which activities are ahead or behind.

THREE-WEEK SCHEDULES

A successful project manager must manage his project on three different planes. He must directly supervise the construction effort underway. He must look at activities scheduled for the next three weeks to ensure an uninterrupted flow of resources to the project, and he must keep an eye on any long lead items which, if not tracked continuously, could eventually cause work stoppage or delay. A sample three-week schedule is included in this chapter. Each day includes the work scheduled for a three-week period, Monday through Saturday. The items of work listed on the three-week schedules must be clear and measurable. The three-week schedules must show the work shown on the Level III bar chart for that period. If you are behind schedule, the three-week schedules must also reflect how you are going to get back on track.

Key resource requirements for the activities scheduled for the next three weeks are listed on this schedule. This tool is used primarily by the project superintendent to ensure all materials required are either on the jobsite or have been requested with sufficient lead time to ensure availability. The three-week schedule will be used in the crew briefings described below and also provide ongoing project status to upper management. Three-week schedules are also referred to as weekly goals since that is exactly what they are.

REQUESTING RESOURCES

Resources are going to be an ongoing concern with any construction project from start to finish. Making sure the resources you need are available at your time of need is much less painful when you adhere to the schedule and pay attention to required lead times, but issues are bound to still arise no matter how well you plan. For starters, if you don't understand what the turnaround time is for a particular item – ask! The more lead time given, the better possibility the material will get there in time. This should all be taken into account with the upcoming three weeks' worth of work in mind, which is a rolling period. You don't handle materials on a Monday morning and then wait three weeks to repeat the process. Every week you need to look an additional week out and plan accordingly.

But then you have the issues that are not easily planned for. Everyone involved in a construction project is human and mistakes are bound to happen. Maybe estimating miscalculated the amount of certain parts needed or the bill of materials did not match the actual amounts shipped to the site. In any of these situations, you might find yourself running out of a particular part or material. The same situation can arise when you have the correct quantities but for whatever reason something breaks or is defective and cannot be used. You as the Project Manager need to have a process in place to address these discoveries in the moment as well as ensuring your subcontractors do as well.

CREW BRIEFINGS

Crew meetings are essential. Workers must understand their tasks and execution methods, but that's just the beginning. They require comprehensive knowledge of safety risks and protective measures, quality standards, and timing requirements. They must be informed about the allocated duration for their current work and understand how delays could affect the overall project timeline. In the construction industry, we use different terminology to describe the various types of discussions we conduct with our work teams:

- **Morning Huddles:** These set the tone for the day and lay out what the expectations are. With the larger schedule in mind, certain objectives must be met every single day and we never want to assume our crews are showing up every morning with a clear idea of where they left off the day before or where we need them to focus today.

- **Toolbox Talks:** The main focal points in these meetings are the tasks you need to focus on to keep the project moving forward. They are very similar to a morning huddle in theory, except these can happen at any point throughout the day. Before going on break or lunch, as soon as you first get back, and even right before packing it in for the day. We want to consistently reinforce who is doing what and why.

- **Safety Briefings:** The name speaks for itself, and safety should be an all-the-time focus, not just after a specific talk. But there will be times on a job site where you will move into new phases of the project where the potential for new hazards will arise and these briefings are crucial to minimizing hazards. For example, your team may not have needed safety harnesses while working on the foundation, but once the building reaches a certain height this will now become a requirement worth reminding them about.

- **Prep Meetings:** These occur before a new phase of work begins, often with new vendors/subs, to lay the groundwork for what is to come next. This could include daily production expectations, delegation of manpower, or even where to store tools and material.

PROGRESS MONITORING

Project management choices heavily depend on our current position regarding project advancement. To assess progress, we compare our actual completion percentage for work-in-place (WIP), explained in more detail in Chapter 9, with the planned completion percentage. We aim to maintain close alignment with the scheduled progress line. We seek to avoid overestimated projections or cases where projects conclude months ahead with half the planned effort. Our focus remains on creating achievable yet ambitious timelines. And we cannot just assume that the job has progressed perfectly in accordance with the schedule as there are often drastic differences between the actual WIP competition percentage and where the schedule suggests we should be.

For a Project Manager to accurately represent what they think the actual WIP completion percentage to be, it is a best practice to walk the jobsite on a regular basis with someone like the Superintendent who has an intimate knowledge of the daily progress. On these inspections the Project Manager should be keeping track of the actual completion they see with their own eyes for each of the subcontractors as well as any self-performed work. These estimates are what will allow the Project Manager to then invoice the Owner or their representative for payment according to the original schedule and agreement.

EFFECTIVE MANPOWER UTILIZATION

Project managers must optimize the utilization of their assigned workforce to ensure timely project completion. Team members are either contributing productively or they are idle. The project manager bears the responsibility of keeping the team engaged in meaningful work. To enhance productivity levels, project managers need to eliminate barriers that hinder efficient work. While many common productivity blockers may appear obvious, failing to implement a robust strategy to address them from the start will create ongoing challenges throughout project execution at the work site.

Getting to the jobsite is one of the first obstacles to contend with on a daily basis. Every morning this can be an obstacle without proper planning. Any tools or materials that need to be picked up should be built into your trip to the jobsite if at all possible. Choose a time and route that minimizes time and optimizes time spent on the jobsite. Require a report time from your Subs and do not accept excuses. They must commit an honest workday everyday – otherwise you will not stay on schedule and you are likely to overrun your budget.

The supervisor and/or project manager will decide how often and how long breaks occur to keep all crew members on the same cadence and in accordance with applicable labor laws and regulations. These determinations should depend on how strenuous the work is, temperature, and other site conditions. Ensure anticipation of breaks does not cause unproductive time on the jobsite. Crew members anticipating lunch or break times can cause major losses of time. Ensure everyone is expected to work until it is break time.

Outside of lunch and other mandated breaks there are always going to be the instances where personal situations may arise. Medical and dental appointments are always big ones because when your crew is working during normal business hours there might be other times where their provider can see them. Whenever possible try getting them to schedule these during their lunch break to avoid losing them for an entire day. It actually works out better for all parties because you keep more of your schedule intact and they do not lose paid time by missing a full day of work.

Administrative issues can also contribute to issues with manpower efficiency, so it is important to establish policies for taking care of business related tasks. A good Project Manager takes care of as many administrative functions for his project supervisor as possible, helping to minimize their absence on the jobsite and keep their focus on the mission. Remember the more you do without help of others, the more productive work your on-site crew is capable of performing.

CHANGE MANAGEMENT

An Owner may desire changes or modifications to the work for any number or reasons. Upon written notification thereof, and the General Contractor's written consent thereto, the General Contractor shall perform such changes. The General Contractor will be compensated on a mutually agreed upon amount for such changes and the contract amount revised to reflect the same. Depending on the impact these changes have on the project schedule, the time for completion might also need to be revised accordingly and for any other cause beyond the reasonable control of the General Contractor as a result of the original change.

When these changes occur prior to the project being awarded or commencing, it is the A/E design manager's responsibility, along with the design team, to prepare amendments to the bidding documents and issue them to the General Contractor. The General Contractor issues the changes to the bidders. Along with the design document changes, the design manager prepares an addenda form. The form has a place for the bidder to acknowledge receipt of the addenda. The General Contractor shall verify all addenda were included in all bids received. All project correspondence between contractors or suppliers is tracked via a log prepared by the Project Administrator.

Before construction starts, the on-site QC Manager shall prepare the Field Test Register form. This form outlines all the formal quality control testing that is required on the project along with the frequency of testing. The designers will outline required testing in the contract documents. The Construction Quality Control Report sheet will be completed daily by the on-site QC Manager. See form in Appendix 9. The Quality Control Organization will perform, at minimum, the Three Phases of Quality Control for each Definable Feature of Work (DFOW). A DFOW is a distinct and identifiable component or process in a construction project. Your DFOW's must be identified and approved in your Quality Control Plan (QCP).

There will also be numerous occasions where changes will be required after the construction project has begun. These can be trickier because of the complexity of changing aspects of construction that have already been completed as opposed to just making changes on drawings and material lists. The financial impact can also be much greater depending on how many others are impacted and whether or not materials subject to change have already been shipped to the job site or potentially installed. For that reason we have multiple steps involved when requesting these changes.

RFI

An RFI is a formal request for information when a discrepancy is found in the course of construction. The issue might be in the project drawings, project specifications, contract drawings, etc. We are asking for clarity from all relevant parties to see what options are available to rectify it. The form will indicate what is going to be impacted and by how much as it pertains to time delays and cost overruns. This is where we start to get to the bottom of who is responsible for the issue. Samples of an RFI (Appendix 8-1) and the RFI submittal log (Appendix 8-2) can be found in the Appendix.

Responsibility

Conversations will take place between the key stakeholders. Generally the Architect and Engineer will talk through this at a high level with the Owner and Project Manager to decide how the relevant parties should proceed. Unless it is a black-and-white issue where we can decisively say who made the error and should bear the responsibility for fixing it, the conversations will generally filter down to the individual trades impacted by the conflict in an attempt at finding the fastest and most cost-effective solution.

Receiving an RFP

Once we know who the impacted parties are, the RFP is where we ask the impacted trades to create a formal scope of work to address the RFI and request the price associated with making the necessary changes.

Pricing an RFP

This is the stage where the impacted trades must determine how much the additional work will cost, which involves verifying quantities, confirming the scope is accurate, and addressing the additional time needed so it can be added into the schedule. General Contractors are not expected to do the work at cost, but they are also not expected to use this as an opportunity to take unfair advantage of a situation. There is an acceptable mark-up range, especially in the federal market, that accounts for both overhead and profits in addition to the material and labor needed to perform the work. Significantly deviating from these prescribed norms can expose your Company to an audit, which can either land you more favorable terms or get you in trouble if you have been charging more than the acceptable norm.

Negotiating Price

This mainly entails verifying that the contractors in question are using the correct quantities and scope of work on the RFPs they send back. Some contractors have a reputation of underbidding jobs and then trying to make the money back on changes so vigilance is recommended. The project budget generally has a balance available for contingencies like this since issues arise more often than not, but it is the job of the Project Manager and the Owner's representative to minimize how much is paid out and avoid cost overruns.

Impacts to Scheduling

Time is money, especially when it comes to delays in a construction project. Depending how significant a change request becomes, it could not only impact the time needed for the impacted contractors to make the adjustments, but those delays can trickle down to everyone else on the job site that becomes delayed as a result of needing to wait for the corrective work to take place. This is where we get into compensable and non-compensable delays. If your specific schedule is impacted as a result of the delays then it is compensable but if your work will still finish on time it is not compensable.

Request for Equitable Adjustment

This is a situation that doesn't happen often, but when it does it can be significant. Several factors can lead to this type of compensation request. One instance can be where the overall project overruns the schedule by such a significant amount of time that the contractors ability to bid new jobs is negatively impacted because their manpower is tied up longer than expected. Another would be where the work winds up getting completed out of the original sequence, which can lead to more mobilization costs or other logistical damages that cost time or money.

Change of Building Occupancy Date (BOD)

During the planning process, the ways any changes in the BOD will be handled must be considered before finalizing a BOD. A BOD is imperative to Customers since the completion of the project is the single factor controlling the usability of the structure. When a delay of the BOD becomes necessary, the Customer must be notified in writing of the cause and the estimated duration of the delay. The BOD is reflected in the Level II SITREP (Situation Report), however, shall not be adjusted until approved by higher authority. This will ensure the project's goals do not differ from the established BOD.

CHAPTER 9

PROJECT MONITORING

Project monitoring refers to the methods used to assess a project's status and compare actual progress against the planned schedule. In order to monitor a project's progress, project managers must be knowledgeable about completing direct labor reports, figuring WIP, updating bar charts, submitting SITREPs, and arranging project photos. This chapter will explain the techniques used to monitor a construction project.

PERSONNEL

Superintendent: As the point person for coordinating everything related to the schedule on the job site, the Superintendent is walking around on a daily basis. They are making sure the right number of tradesmen for each subcontractor are on the job site and working on the areas most crucial to the schedule. They know down to the day what needs to be accomplished for every trade and are responsible for catching miscues as soon as possible. They will also handle any mishaps, setbacks, change orders, and anything else that could become a liability for the schedule, financial loss, or injury.

Quality Control Manager (QCM): The Quality Control Manager is looking specifically at the work being put in place, ensuring all materials meet specifications and are being installed in the correct locations. They also ensure minimum timeframes are being adhered to for certain types of installation such as concrete curing.

Security: Most job sites will have 24/7 security on-site to prevent theft, vandalism, or unauthorized entry. During normal business hours where work is conducted they will mainly ensure only authorized personnel are coming and going and that no equipment or materials leave with them. Their responsibility is the same after hours but with less activity and people around to act as deterrents, they must be more proactive in conducting regular rounds to preserve the integrity of the site.

Site Safety and Health Officer (SSHO): Even though the safety officer has nothing to do with the project schedule, their vigilance is very much required to keep everyone safe. Routine inspections of the jobsite ensure all safety gear and harnesses are in working order, no work is conducted in an unsafe manner, there is no alcohol or drug consumption while on the clock, and all other instances where someone could wind up hurt.

ADHERENCE TO SCHEDULE

Once you set foot on-site and develop your schedule, there is no deviation. If the scheduled completion date becomes unattainable at any point, upper management and the Customer must be alerted to this problem. It is important to keep documentation on delays, outlining exactly what contributed to it so you can explain yourself in an informed manner.

REGAINING THE SCHEDULE

Many unanticipated problems crop up in the execution of our tasking. All these problems will likely have at least some impact on the schedule. Finding ourselves one or two percent behind schedule is not a catastrophe, but we need to formulate a plan for getting back on track, for regaining our schedule. There are several strategies for regaining your project schedule and we are going to look at a few.

Better Methods

Frequently, there are quicker construction techniques available than those we initially planned or are presently using. When falling behind schedule, reviewing the CASS for future tasks could uncover chances to reduce man-days through modified methods. Typically, superior equipment leads to decreased time expenditure. For instance, in an underground pipeline project, utilizing a trencher instead of a backhoe could reduce man-days, assuming the budget permits. Never hesitate to seek input from colleagues regarding man-day-saving options. This exemplifies the principle of "working efficiently, not excessively." The most negative response you might receive is "no."

Increase Effective Workday

Obviously getting up earlier each day to go to work is a way to extend the workday, however selling that to your Subs may not be an option. Increasing the availability in any way will help. Cut lunch breaks from 45 minutes to 30 minutes, ask subcontractors to move to a closer living location to site, and numerous other items can make your day more effective. For planning purposes we use between a .95 to a .85 availability factor. The actual calculation is in the formula below:

$$AF = MD \text{ (expended)} / (\text{Crew (assigned)} \times WD \times MDE)$$

To calculate the actual availability factor, we need to know the crew size, the total man-days worked over a specific period, and the number of workdays within that timeframe. If the actual availability is low (below .85) you want to consider a permanent change to the daily routine to increase your availability factor. Work with your subcontractors to increase your availability factor, which will in turn decrease the number of days on-site.

Phasing of Activities

Most project plans begin with a logical sequence diagram primarily utilizing finish-to-start dependency connections. This has the effect of stretching the project duration and reducing the crew sizes. It also leaves plenty of opportunity to compress the schedule by working several activities at the same time. It's possible to save several days from your timeline by dividing your team and assigning a portion to begin the subsequent task. To achieve meaningful progress on your schedule, it's likely that additional personnel will be required.

DAILY PRODUCTION REPORTS

Production reports are the most accurate way to record man-days and overall duration in calendar days being expended on a construction project. These systems help track your team's performance and responsibility at work. They are the basis of your SITREP, therefore it is imperative the labor report be filled out correctly. Project Managers are responsible for reviewing the daily production reports from the supervisors in the field. Each day these reports should reflect man-days and duration expended by personnel assigned. So if you have a crew of five personnel each day, you should log five man-days, with one duration/calendar day per task. The field supervisor must file these reports in the project folder daily.

The content on the daily production reports should consist of:

1) **Productive Labor:** The number of man-days expended directly contributing to the completion of the job. These man-days are subsequently divided among activities that are being performed.

2) **Indirect Labor:** The number of man-days expended to support construction operations, which do not ultimately reflect on the WIP.

3) **Training:** The number of man-days expended for schools, training, and any other time spent off the jobsite learning.

4) **Overhead Labor:** Simply put, overhead labor is you, the Project Managers, along with engineering and the business development staff. As a project manager, every moment of your day should be spent supporting a project. That is authentic overhead. Any time not spent supporting your projects does not fall into this category or any other for that matter because it is nothing more than wasted time.

SITUATION REPORT (SITREP)

A SITREP should be developed and reviewed weekly by management. Once SITREPS are developed they should not be closed out till the work is 100% complete, including punch list items. Once the close-out report has been completed, the project is considered complete and the close-out SITREP will be developed for final analysis during the project debriefing. To compile a SITREP, the Project Manager should take the input from the daily reports and put it into a usable format every week. This format is known as a Level II SITREP and the Project Progress Report (PPR).

The Level II SITREP is a graphical description of your project's status. In addition to the graphic are the various analytical tools based on calculations of scheduled man-days versus work-in-place (WIP) man-days. The PPR gives the field supervisor the discretion to determine what percentage complete he is on each section of work. So if he has to determine the percent of WIP for an activity like framing, he must determine what is considered 100% and evaluate his current status. Round figures like 10, 25, 50, 75, 90, 100 are common percentages used by field supervisors. Each activity individual percentage complete is given a weighted average and combined to give an entire project WIP.

All Level II SITREPs shall include a summary, which provides project work-in-place, remaining project man-days, and completion date percentage data, as well as a short narrative of the work accomplished during the reporting period for each project. For projects where no work was accomplished during the reporting period, the Level II SITREP will reflect no change in percentages. If no tasks are completed, a clear explanation must be provided in the comments section to justify the lack of progress.

- **Weighted Percent:** The man-days estimated for the activity divided by the total project man-days.

- **Activity Work-in-Place (WIP):** The percent of WIP per activity. Each activity must be evaluated daily on the WIP; weekly this WIP should be displayed on a SITREP to upper management. On a daily basis, the daily report will tell both the project manager and the field supervisor where they are at for work-in-place and where they should be. This helps the field supervisor make key decisions on how long each workday and work week is. This also will allow the field supervisor to see the loss or gain of productivity. This daily tracking will empower your field supervisor, while allowing you as a project manager to manage the job effectively and keep the crews focused on productive use of the workday.

- **Project % Complete:** The percentage the work completed on that activity contributes to the overall project completion. We get the project % complete by multiplying the weighted percent by the activity % complete (WIP) for each activity.

- **Actual Percent Complete:** The total of the project % complete column. It is important to compare the actual to the scheduled WIP % to determine the status of your project. Evaluating factors like man-days expended will help you determine the best corrective action to take in order to keep on schedule.

- **Man-days Remaining:** A reflection of how much work remains to be done on the project and has nothing to do with how many man-days have been expended. Once an activity is complete the man-days remaining should not be considered as float time. These activity man-days were given from a particular activity not the project as a whole. Each activity should have an activity man-day estimate that is the driving goal per activity. If the WIP is 50% on a 10 man-day activity the remaining man-days should be 5 regardless of the amount expended.

- **Man-days Expended:** These have nothing to do with % complete. This information is included on the production report to give insight to management where work is ongoing and how the work has impacted the schedule. The total man-days expended is also needed to determine the scheduled completion to use, to compare the actual work-in-place.

- **Comments Line:** The section is on the Level II SITREP and the daily reports used to explain the circumstances for why a project has fallen behind and also to determine the corrective actions being taken to rectify the schedule. Delays, resolutions, manpower issues, and many other topics should be covered in this section.

BAR CHARTS

Bar charts help visualize and monitor your project's advancement over time. They indicate whether you're moving faster, slower, or according to plan. At the end of each week you should plot your bar chart to determine if the bar chart actually reflects the intended work schedule. The bar chart is a good visual way of laying out the course of a project against other projects that fits into the same timeframe.

PHOTOGRAPHIC COVERAGE

Project managers shall provide at least ten color photographs of each active project on a weekly basis. Vantage points providing the broadest coverage should be chosen, and the same view should be utilized for PowerPoint presentations to show the sequences of construction. Also include pictures of the crew working and any major construction evolutions depicting prominent work. It is important to ensure no safety violations are depicted in these photographs, so please review your pictures before forwarding them to any higher authority.

PAYMENT

The General Contractor shall submit invoices, with appropriate detail, monthly as the work progresses. The Owner shall pay the General Contractor within ten to thirty days of receipt of such invoices. Disputed amounts shall not preclude payment of undisputed amounts. Amounts paid after 30 days shall be subject to a pre-disclosed monthly interest rate, generally one percent per month on all outstanding balances. Payment methods shall generally consist of checks, wire transfers, or ACH transfers. Cash payments are never advisable.

FORCE MAJEURE

Delays or defaults in performance under an agreement (other than the payment of money) shall not constitute a default of a party's obligations hereunder if the same results from or is caused by occurrences beyond the reasonable control of the defaulting party, including but not limited to: changes ordered in the Work and other effects of acts or omissions by Owner; acts or omissions of other contractors whether or not employed by Owner; acts of God; storms; floods, fires, explosions, or other casualty losses; unusual weather conditions; strikes, boycotts, lockouts, or other labor disputes; unavoidable delays in transportation and delivery of material and equipment; and acts or omissions of the public enemy or any Client, its agencies or officers, federal, state, or local. The General Contractor shall be entitled to an adjustment in schedule and compensation due to such occurrences.

PROCUREMENTS

The Project Manager prepares a requisition in such a way that a buyer can prepare purchase orders for goods and services to meet the plans and specifications. The items, which are purchased for projects and tracked via logs, are prepared by the Assistant Project Manager and issued to the Project Manager weekly. The buyer tracks the status of the purchase as necessary to assure delivery time is met.

CONTROL OF ON-SITE CONSTRUCTION

The Quality Control Manager will perform sufficient control phases and tests of all work, including that of contractors, to ensure conformance to applicable specifications and drawings with respect to the materials, workmanship, construction, finish, functional performance, and identification.

RECORDING FORMS

The quality personnel will perform all tests as indicated in the contract specifications using the appropriate American Society of Testing and Materials (ASTM), or other approved standard test methods. The following list itemizes some of the forms, which the contractor quality control personnel intend to use. This list is not all-inclusive and may be revised and updated as conditions require. The contractor's records will be available for review by the Client upon request.

Contractor Daily QC Report: To be filled out daily by the contractor's quality control personnel covering the day's quality control activities, approved by the General Contractor's Quality Control Manager, and placed in the General Contractor's file.

Construction Quality Control Daily Report: To be used by the Quality Control Manager to report the day's quality control activities of the General Contractor and all contractors, submitted to the Client daily.

Trip Report: Used to report activities covering offsite trips and will be submitted with the contractor's Quality Control Daily Report

Correspondence Log: All project correspondence between contractors or suppliers is tracked by the project administrator.

CHAPTER 10

CHANGE MANAGEMENT

In federal construction, change is not the exception — it's part of the process. Understanding how to manage change orders is essential for project success and survival.

WHAT ARE CHANGE ORDERS?

A change order is a formal modification to an existing contract. In federal construction, it alters one or more of the following:

- The scope of work.
- The contract price.
- The schedule or performance period of the contract.

Change orders can result from unforeseen conditions, design errors or omissions, differing site conditions, inconsistent interpretations of contract documents, or directives from the government.

It's important to recognize that in federal contracting, only the Contracting Officer has the legal authority to modify the contract. Informal agreements or verbal directions from other government representatives — no matter how well-intentioned — are not contractually binding. Proceeding with change order work without an approved modification is done at the risk of the General Contractor and their subcontractors.

CHANGE ORDER LIFECYCLE: FROM REQUEST FOR INFORMATION (RFI) TO APPROVED CONTRACT MODIFICATION

Managing change orders requires understanding the lifecycle of a typical change order. It generally unfolds in three key steps:

1. Request for Information (RFI)

The contractor identifies a discrepancy, unclear direction, or potential change in the field and submits an RFI to the Contracting Officer or the architect/engineer of record. The purpose of the RFI is generally to clarify intent – not necessarily to initiate a change. That said, RFIs often uncover information that leads to contract modifications. It is advised to maintain an RFI Log for your project. We detail the necessary information required for the RFI Log below.

2. Request for Proposal (RFP)

If the response to the RFI reveals that the scope must change, the Contracting Officer issues an RFP to the contractor. The RFP will be a formal document that details the scope of the proposed change and any additional information needed to properly estimate the proposed change. This is an invitation to propose a cost and time for implementing the change. As the General Contractor, your responsibilities to the RFP include:

1) Thoroughly review the RFP. Make sure that the scope detailed within the RFP properly encompasses the discrepancy identified in the RFI.
2) Prepare a full cost estimate of the scope of the RFP.
3) Conduct a schedule impact analysis of the proposed change. If the proposed changes impact the Critical Path, you will need to request additional time and/or a revised Contract Completion Date (CCD).
4) Submit a formal proposal with cost breakdown, justification, and Time Impact Analysis (TIA), if applicable.

3. Change Order or Modification (Mod)

Upon submission of your RFP, the Contracting Officer will perform a thorough review. There may be several rounds of questions, clarifications, or negotiations. If the Contracting Officer accepts the proposal (or a negotiated version), the Contracting Officer issues a bilateral or unilateral modification to the contract.

- Bilateral Modification (Supplemental Agreement): Both parties sign — preferred method.
- Unilateral Modification (Change Order): Contracting Officer directs the change without contractor agreement — often used to prevent delay.

EVALUATING TIME IMPACTS

Every change order proposal should assess whether the change affects the project's critical path. This is done through a Time Impact Analysis (TIA) — a forensic schedule analysis that measures the impact of the proposed change using CPM (Critical Path Method) scheduling techniques. Your TIA and schedule analysis will need to follow a recommended industry standard for delay claims analysis. At minimum, your TIA will need to include descriptions of Fragnet activities, narrative formats of the changes to the schedule, type of schedule analysis performed, impact prior to and after insertion of the delays, and justification for the added Fragnets, logic, durations, resources, etc. Putting together TIAs is a very complex task and typically performed by senior level Project Managers or scheduling consultants.

Why does a properly put together TIA matter? Remember, time = money. The Contracting Officer may grant compensable (cost) and excusable (time) delays. However, without schedule substantiation, time extensions are often denied. As a recommended best practice, you always want to include a narrative, impacted schedule fragment, and comparison to baseline.

MAINTAINING A CHANGE ORDER LOG

Tracking change orders is a cornerstone of project controls. A well-maintained log helps manage scope creep, monitor trends, and provide transparency to stakeholders. Your Change Order Log should include:

- Change Order Number
- Description
- Origin (RFI #, directive, etc.)
- Cost estimate
- Time impact (days)
- Status (Draft / Pending / Approved / Rejected)

- Date of approval/modification number
- Notes and responsible party
- Pro tip: Maintain consistency with USACE, NAVFAC, or GSA reporting formats where applicable

UNDERSTANDING GOVERNMENT FUNDING CONSTRAINTS

One of the most unique aspects of federal construction is that even when a change is valid, funding availability governs whether it can proceed. Simply put, the change order/modification process with the Federal Government can be cumbersome and time consuming. Always account for the necessary time needed for your Contracting Officer to review, obtain approval, and issue your modification. This process can take a few days up to a few months.

IMPORTANT POINTS

Each modification must be backed by appropriated funds — the Contracting Officer cannot exceed the limits of allocated funding. The government may acknowledge the validity of a change, but defer action until funding is available. Some agencies (e.g. USACE) issue Unpriced Change Orders to keep the project moving while cost/ funding negotiations continue. Remember, proceeding with work without an approved Change Order is often at your own risk. As the General Contractor, you should monitor funding obligations closely, track unfunded approved change orders separately from funded ones, and push for prompt negotiations with the Contracting Officer to avoid placing cash flow strain on your company and project. Push for prompt negotiations to avoid cash flow strain.

COMMON PITFALLS AND BEST PRACTICES

Pitfalls

- Proceeding with work based on verbal direction.
- Failing to submit proposals timely.
- Not analyzing or documenting schedule impacts.
- Overlooking minor changes that cumulatively affect cost or time.

Best Practices

- Treat all changes — even "small" ones — as formal contract issues. Don't be afraid to issue "no cost, no time impact" change orders. These can be excellent bargain chips in later negotiations.
- Train field staff to document potential changes clearly and consistently.
- Keep a paper trail: RFIs, meeting minutes, directives, and submittals matter. This is especially important if work has proceeded on an unapproved Change Order.
- Engage project controls early in the process — don't wait for accounting to catch up.

STRATEGIC ADVICE FOR NEWCOMERS

- Understand that change orders are not adversarial — they're part of the contract's built-in flexibility.
- Build a culture of documentation — it's your best defense and strongest negotiation tool.

- Maintain open, professional communication with the government team — trust accelerates resolution.
- Track both cost and time together — too often, one is emphasized at the expense of the other.
- A very important concept to consider when negotiating your time associated with a Change Order. Total Float belongs to the project. It does not belong to the Government or the General Contractor.

Change orders are the lifeblood — and sometimes the headache — of federal construction. Mastering the process, documentation, and strategy behind them is essential for effective project controls. As a project control professional, your job is not just to respond to changes — it's to anticipate them, manage them, and turn them into successful outcomes for your organization and the government client.

CHAPTER 11

SAFETY

The Company's Safety Officer is not responsible for your jobsite safety. The Owner and the designer are not responsible for safety procedures. YOU, the Project Manager, are solely responsible. The Safety Officer is responsible for oversight and technical guidance. If you have any questions concerning safety on your project, consult the Safety Officer. Safey on a construction project is everyone's responsibility and ignorance is no excuse. It is your responsibility to do construction in a safe manner no matter what obstacles might arise and also to hold all your subcontractors to the highest possible safety standards.

Good housekeeping is an integral part of any safety program. As best practice, the General Contractor shall clean up trash and debris created by its work and personnel daily. Scrap material shall be hauled out of the facility daily and disposed of by the General Contractor. The General Contractor shall remove from the job site all combustible scrap materials daily. All material shall be stored in an orderly manner and kept clean. Should the General Contractor's housekeeping not meet the standards defined herein, after 24 hours advance notice, the General Contractor shall be back-charged for the cost of cleaning its work area.

The goal of a safety program is to prevent mishaps, and sometimes housekeeping alone is not enough. We avoid using the term "accident" because it suggests a lack of responsibility, as if the incident occurred without cause or fault. Mishaps most commonly result from the failure to follow safe construction practices. These procedures should supplement the Project Specifications Section of Safety and Health. They shall be considered additive to all applicable federal, state, and local safety requirements. However, should any of these requirements be in conflict with any requirement of such other applicable requirements, the most stringent requirements shall apply.

When in doubt, you should always refer back to the approved Accident Prevention Plan (APP) that was outlined in Chapter 4. If something is unclear the next step would be to use what we consider to be the safety bibles in construction, the 29 CFR and EM385, but here are some best practice process for avoiding mishaps on the job site:

1) **Recognize Hazard:** We begin by recognizing that construction is a dangerous business. There is a potential for death or serious injury on our job sites on a daily basis. Identify very specifically what the hazard is that could cause the death or injury.

2) **Identify Corrective Action:** Our primary reference for preventive measures is the Code of Federal Regulations (29 CFR) and the EM385. This guide holds all the information

to determine the safety requirements needed when doing any type of construction work. The difficult thing is to identify the hazards your crew will come in contact with and the governing guidance to utilize, remembering the stricter applies. Once you have identified these items, the 29 CFR or EM385 has the information you need.

3) **Obtain Equipment and Material Training:** The Project Manager and Safety Representative will provide assistance in obtaining all materials, equipment, and training for safety related items. The safety equipment should be included in the project budget and bought for each person assigned to an individual project. Resource management should be integral to setting up safety training, with proper coordination of the Project Managers.

4) **Ensure Personnel Awareness:** Use the daily five-minute safety lectures (Tool-Box Lecture) to ensure the crew understands the proper use and wear of safety equipment provided and the locations of this equipment. Safety lectures must address all hazards identified on the CASS for work scheduled that day. Also inspection of safety gear after each safety lecture is a great practice to start.

5) **Proper Supervision:** The Project Manager is responsible for ensuring the crew and all other personnel use proper protective equipment at all times.

6) **Emergency Response:** To ensure an emergency response is not delayed in the event of a mishap, the location of the nearest phone, map showing nearest medical facility, or first

aid station, and emergency phone numbers must be posted on the jobsite.

7) **Investigate and Report:** Any mishap (regardless of how minor) or near miss must be investigated and documented to minimize the chance of it happening again. A form for the supervisor's report of injury is provided as Appendix 10-1.

VIOLATIONS

The following violations are grounds for immediate discharge of General Contractor personnel:

- Gross disregard, refusal to obey, or repeated violations of safety and health rules and regulations
- Fighting (physical contact), horseplay, gambling, or sleeping on the job
- Theft or other illegal conduct
- Possession or use of alcohol or illegal drugs
- Willful destruction of property
- Tobacco use outside of designated areas

BEST PRACTICES

A representative of the General Contractor must be on the site when any work is being performed by a lower tier General Contractor, even if no work is being performed by the General Contractor's direct work forces. These representatives shall be in charge of general coordination with the contractor and shall be responsible for assuring that all safety regulations are observed. In addition, they shall be responsible for responding to medical emergencies related to the General Contractor's employees or its contractors. Some additional common sense best practices are:

- All construction and employee vehicles are to travel at a speed of no more than 10-15 MPH and shall proceed cautiously in and around the job site and obey all posted speed limits.
- No firearms or explosives are allowed on the job site.
- No glass bottles are allowed on the job site.
- No hazardous waste of any type shall be disposed of on the job site.

SAFETY ITEMS REQUIRED ON THE JOBSITE

The following safety equipment is required on all project sites. See the 29 CFR and EM385 for additional information.

- **Emergency Plans:** Each jobsite must have posted the location of the nearest phone, the telephone numbers, and reporting instructions for ambulance, hospital, physician, police, and fire department personnel.
- **First Aid/CPR Qualified Personnel:** If a medical facility is not readily accessible (due to time or distance) two crew members must be first aid and CPR qualified.
- **First Aid Kit:** One for every 25 personnel on the jobsite and must be inventoried weekly.
- **Toilet Facilities:** If toilet facilities are not readily available portable facilities must be provided.

- **Drinking Water:** Water must be provided from an approved source and labeled for drinking only and is not to be used for other purposes. Sharing of common cups is not allowed.
- **Temporary Fencing:** Is required if the area is actively utilized by the public.
- **Warning Signs:** Red for immediate hazard, yellow for possible hazard.
- **Eyewash Facility:** Required where personnel are exposed to or handling hazardous substances.
- **Fire Extinguisher:** Placement requirements vary depending on the type of extinguisher and applicable local or state codes. As a general rule, provide at least one extinguisher per floor and ensure coverage based on square footage in accordance with the most stringent requirement — whether dictated by code or by the extinguisher type.
- **Material Safety Data Sheets:** For all hazardous material on the jobsite.
- **Safety Manual:** 29 CFR and/or EM385 are required to be kept on the jobsite for ready access.

PERSONAL PROTECTIVE CLOTHING AND EQUIPMENT

Personal protective clothing and safety equipment are the foundation of any safety program. No matter how careful you try to be on a job site, the hazards are everywhere and the potential to get injured is too great. Even when wearing all the proper gear, injuries can still occur, which is why our goal is to minimize the extent of those injuries. Some of the items in this section might seem like common sense and others might seem unnecessary, but they are on the list because at one point in time a significant injury arose as a result of non-compliance. Here is a comprehensive, but not exhaustive, list of practices everyone on your job site should be aware of:

- **Hardhats:** The wearing of nonconductive hard hats meeting the requirements of ANSI/ISEA Z89.1 is required at all times, except in break areas, offices, or restrooms. Hard hats shall be worn with the bill in front to provide protection against falling objects.
- **Footwear:** Heavy protective shoes shall be worn at all times during the course of all construction activities. Tennis shoes, track shoes, sneakers, and all other footwear are prohibited.
- **Eyes:** The wearing of safety glasses with side shields meeting the requirements of ANSI/ISEA Z87.1 is required at all times on the job site – except while in break areas, offices, and restrooms. Full-face shields shall be worn over safety glasses for all grinding operations, sawing with abrasive cutting blades, overhead drilling, in cutting with a hacksaw or bandsaw, and entry into process lines.
- **Upper Body:** Tank tops, low-cut shirts, or sleeveless shirts are prohibited. Loose fitting garments, shirttails, floppy sleeves, or loose jewelry that may catch or become entangled in equipment are prohibited. Employees shall be sent home to change clothing if they come to work improperly dressed.
- **Hair:** Lengths below the nape of the neck shall be put up under the hard hat or contained in a hair net. Ponytails are not considered proper containment of long hair.
- **Legs:** Long pants are required at all times.
- **Equipment:** OSHA approved hearing (ANSI S3.19-197 and ANSI/ISEA S12.6-2016) and respiratory (ANSI/ASSE Z88.2)

equipment shall be worn when required. The General Contractor shall meet the selection, fitting, and maintenance requirements of OSHA.

PROJECT MANAGER'S RESPONSIBILITY

The project manager's responsibility extends to everyone they supervise. The project manager is the key person in an aggressive safety program. His knowledge shall include, but is not limited to:

- Understanding and following workplace safety protocols and guidelines for designated areas while maintaining appropriate precautionary behavior at all times.
- Enforcing safety rules and correcting unsafe acts.
- Inspecting job and work sites for hazards and corrective actions.
- Educating and training personnel in safe work procedures and rules.
- Reporting all mishaps and near-mishaps to the safety office promptly and ensuring personnel receive immediate medical treatment.
- Investigating all mishaps, determining the root cause and taking corrective action.
- Reviewing safety and health records on crew members.
- Implementing remedial measures for safety concerns identified by workers without penalizing those who report them.
- Supplying appropriate safety gear to staff members and ensuring they properly use and care for the equipment.
- Seeking guidance and support from the Safety Department for effective execution of the OSHA program.

- Understanding workforce limitations and refraining from assigning dangerous tasks to individuals who lack the physical or mental capacity to perform work safely.
- Taking faulty equipment, materials, or tools out of operation until necessary repairs ensure safe functionality.
- Installing relevant safety warning signs in visible locations near or on equipment, storage spaces, materials, and other identified hazardous zones.

CREWMEMBER'S RESPONSIBILITY

Project Managers must verify all personnel understand their duties as outlined below:

- Being familiar with, comprehending, and following safety protocols and regulations relevant to their designated work or workspace.
- Arriving at work well-rested and mentally prepared to perform assigned duties.
- Comprehending and following health and safety measures applicable to their tasks or work environment.
- Notifying immediate supervisors about unsafe conditions including emerging or unusual hazards or potentially dangerous materials.
- Warning others who might be at risk from suspected, known, unusual, or developing dangers.
- Informing direct supervisor about any accidents, injuries, or signs of health issues occurring during work activities.
- Utilizing all required, approved, and provided protective gear and clothing necessary for safe task completion.

- Wearing work-appropriate attire for assigned duties.
- Necklaces and flowing scarves must be avoided when they pose a safety risk to the wearer.
- Keeping hair and facial hair properly contained when they could pose risks near equipment or flames or obstruct clear sight.

HAZARD COMMUNICATION

The General Contractor shall comply with OSHA Hazard Communication Standard 1910.1200 and all state and local hazard communication requirements, which shall include but not necessarily limited to the following:

- Develop and implement a written Hazard Communication Program for the site. The General Contractor shall provide a copy of the Program to the Contracting Officer.
- Train its employees in handling all hazardous materials present in the workplace.
- Maintain a list of all hazardous materials present in the workplace and post it in a place accessible to all employees. Safety Data Sheets (SDS) must be on file in the General Contractor's site office for all such materials. The SDS shall be furnished to all personnel on the jobsite, including contractors and subcontractors, who may come into contact with or be affected by hazardous materials.
- The General Contractor shall assume responsibility for the cost of obtaining all permits required by the appropriate governing agency, associated with the use or disposal of any hazardous waste material generated during the performance of its work.

- The General Contractor shall conform to the provision of CFR 1926.59, Hazard Communication Program.

SAFETY MEETINGS

Safety can never be taken too seriously, which is why we always want to make sure all personnel are on the same page when it comes to reducing hazards and injuries. When we covered Crew Briefings in Chapter 8, we spoke about some of the meetings we should be having in this regard, which included weekly site safety, tool box talks, and stand up safety lectures (huddles). But there are a few more specific to safety that cannot be overlooked.

Weekly Accident Summary

The subcontractor shall complete and submit a Weekly Accident Summary to the General Contractor each Monday morning. A copy of any injury or incident report shall be forwarded immediately including Employer's Report of Occupational Injury or Disease to General Contractor's site Project Manager. The General Contractor will furnish these reports to the Client on a weekly basis.

OSHA Safety Training

The General Contractor at all times shall comply with all of the safety rules and regulations established for this project, as well as all applicable provisions of the Occupational Safety and Health Administration 29 CFR, PART 1926 "Safety and Health Regulations for Construction," and 29 CFR, PART 1910 "Occupational Safety and Health Standards for General Industry" and as may be amended from time to time, and all other federal, state and local regulations as they may affect the work on the project. The General Contractor shall comply with all amendments to the aforementioned rules and regulations immediately as they become effective. The OSHA thirty-hour safety training should be a mandatory requirement for all project supervision personnel and when a General Contractor's workforce exceeds twenty-five or more employees, a full-time dedicated site safety person shall be assigned.

Prior to the beginning construction work, the General Contractor shall be required to attend a Project Safety Orientation conducted by General Contractor's Project Safety Coordinator. The General Contractor shall conduct its own employee safety orientation program for all employees based on its Safety and Health Plan developed for this project. Violation of this plan, OSHA, or site regulations, following a single written warning, is cause for dismissal. The General Contractor is also encouraged to provide a means of recognizing safety accomplishments among individuals and groups. The intent is to recognize positive contributions to safety and reinforce the contractor's commitment to it. Any changes in the program, subsequent to evaluation, shall be submitted for review prior to implementation.

ELECTRICAL SAFETY

Every electrical power source on construction sites, including pre-existing outlets in renovation projects, is classified as temporary. All such power sources require inspection, safety certification, and labeling showing the inspector details, Company information, and inspection date before initial usage. Generally, recertifications of jobsite electrical supplies are required bi-weekly or monthly.

Ground Fault Circuit Interrupters (GFCI) shall be used with all power tools, whether double insulated or not. All portable tools, extension cords, small gasoline, pneumatic, and powder actuated tools shall be inspected monthly. Equipment and circuits that are de-energized shall be rendered inoperative and have tags attached at all points where such equipment or circuit can be energized. Refer to lock-out/tag-out procedures.

Each General Contractor shall establish a quarterly assured grounding inspection program of all welding leads, electrical cords, tools, etc. As a best practice, these inspections should be color-coded so that all personnel can easily identify what has been inspected within the last three months and exercise additional caution with those items that have not. A sample color coding system might look like:

- January - Yellow
- February - Yellow
- March - Yellow
- April - White
- May - White
- June - White
- July - Orange
- August - Orange
- September - Orange
- October - Red
- November - Red
- December - Red

Ground Fault Interrupters (GFI's) are required for all receptacles at the power source.

Each General Contractor shall develop a lockout and tag out procedure, to ensure the safety of those working near any potential live electrical sources. The panel or transformer is to be physically locked by the contractors performing the work and a tag with their identifying information placed on the lock for all to know who must unlock the power source and prevent any accidental electrocutions. This procedure shall be subject to the review of the Contracting Officer. It shall be the responsibility of the General Contractor to strictly adhere to this procedure. The General Contractor will monitor the subcontractor's compliance on a frequent basis. The General Contractor will lockout with the subcontractor, as first on/last off.

HAZARDOUS MATERIALS AND HAZARDOUS WASTE

Project Managers need to understand the risks that dangerous substances present to all workers at the construction site and implement safeguards. Let's explore fundamental protocols established to reduce accident and injury potential.

When purchasing any hazardous material it should come with a SDS, but if for some reason it does not, you must obtain the SDS. The manufacturers of potentially hazardous chemicals, cleaners, solvents, oils, etc., will outline what precautions should be taken in handling, storage, use, and the means of first aid that should be rendered in the event of exposure to that specific material. It will also identify any personal protective equipment for the safety precautions required as well as first aid or medical treatment required for exposure. The foreman is required by federal law to inform his crew of the risks and all safety precautions associated with any hazardous material present on the jobsite. This should be done during each daily safety lecture. Additionally the SDS must be posted conspicuously on the jobsite.

Hazardous Material Storage

The safest practice concerning hazardous material is not to buy more than you can use in a week. Storing hazardous materials on the jobsite requires the use of approved storage containers. Consult your safety office for hazardous material storage.

Asbestos Operations

We do not touch, move, or come in contact in any way with potential asbestos materials. If asbestos is present, contact a certified asbestos contractor, the Project Manager, and the safety representatives.

Respirator Protection

Before respirators can be utilized, these essential conditions must be fulfilled:

- Proper equipment selection by safety department
- Health assessment of intended users
- Fit testing conducted by qualified staff
- Respiratory safety instruction for all intended users
- Written SOP must be developed for the worksite, including emergency and rescue guidance, and posted on the jobsite

Hazardous Waste Disposal

The disposal requirements will differ greatly based on the type of material, quantity, etc., which makes it imperative to familiarize yourself with the specifics for your job site and the waste generated.

At a bare minimum, anyone responsible for disposing of this waste should know where the disposal locations can be found, the manner in which it must be disposed, and the maximum allowable quantities. This information will be further documented in your Environmental Protection Plan (EPP), Waste Management Plan, or Hazardous Materials Plan. Hard copies should be kept in an easy accessible location on the jobsite (job trailer, superintendent vehicle, etc.).

EXTERIOR SAFETY

While there are very serious and in depth concerns when it comes to exterior safety that we will go into detail on during this section, there are also a number of broader concerns every general contractor should be aware of. The first and most unpredictable is the weather, which is why awareness is so important. Different weather conditions can bring about different conditions and hazards like lightning, flooding, ice, and heat stroke.

Slips, trips, and falls might be the most common category of injury on the exterior of a job site. No tools, stray material or ladders should be left lying around or unattended. Any spots where the ground is uneven or unstable should be clearly marked for all to see or remedied when possible. The same can be said for any unstable structures where workers run the risk of losing their balance and falling.

Traffic hazards are another notable concern and any areas with significant traffic should have personnel in place to direct the flow and ensure pedestrian awareness. Noise control can greatly reduce injuries by allowing everyone in a reasonable range to hear what is going on around them and better detect and assess dangerous situations. We also want to remain vigilant to dangers from above, like suspended loads or objects falling from the structure, which is why PPE must be worn at all times regardless of whether someone is performing work at the time or not.

Any exterior activities with the potential to cause breathing issues should also be taken very seriously. On the mild side is the danger from dirt and dust sediment in the air associated with the excavation and removal of dirt and debris. On the more severe side of the scale are extremely toxic substances like crystal silica that is created when materials containing silica are cut, crushed, drilled, or ground. This can happen during activities like sawing brick or concrete, sanding, or manufacturing concrete blocks. Whenever performing activities where any of these conditions may exist, refer to the section on respirator protection to ensure you are taking appropriate precautions to prevent long term health issues for workers.

Excavations and Trenching

Prior to opening any excavation or trench, the General Contractor shall provide notification to the appropriate authority. In addition, each General Contractor shall contact any other necessary personnel to determine the location of any underground installations such as sewer, telephone, fuel, electric lines, etc. All excavations shall comply with all OSHA, state, and local regulations. All trenches and excavations shall be properly barricaded and covered 5' back from the edge of the excavation to prevent persons and equipment from falling into the excavations. All walkways or ramps crossing over excavations shall be securely fastened and equipped with standard

OSHA guardrails. A competent General Contractor employee shall inspect excavations and trenches daily. A written report shall be submitted detailing the results of the inspection.

Shoring and bracing shall be set at sides of deep pits or trenches to prevent the possibility of cave-ins. Ladders shall be placed in excavations 4' or more in depth. All banks 5' high or more shall be sloped to the angle of repose (the greatest angle above the horizontal plane at which a material will be without sliding) or shall be adequately shored in accordance with OSHA requirements. Ladders or steps shall be provided in all trenches 4' or more in depth. Ladders or steps shall be located to require no more than 25' of lateral travel before having access or egress. Sloping or benching for excavations greater than 20' deep shall be designed by a registered professional engineer hired by the General Contractor.

All material excavated shall be stored at least 3' from the edge of the excavation or trench and shall be barricaded to prevent material from falling into the excavation.

Scaffolding

Work requiring the lifting of heavy materials or substantial exertion that cannot be done safely from the ground must be done from scaffolding and not ladders. All scaffolding shall be inspected by a qualified individual and approved by the General Contractor prior to use and shall be in accordance with OSHA regulations and 29 CFR. All scaffold planking shall be free of knots and cracks and shall completely cover the work platform. Scaffold planks shall be placed tightly, cleated at both ends or overlapped a minimum of 12" and nailed or bolted to prevent movement. Overlaps shall only occur directly above scaffold supports and nails must be driven full length. No duplex nails are allowed.

No scaffold shall be erected, moved, dismantled, or altered except under the supervision of competent persons and any part of a scaffold weakened or damaged shall be repaired or replaced immediately. All scaffolds shall be erected plumb with footings or anchorage sound, rigid, and capable of carrying the maximum intended load without settling or displacements. This means no unstable objects such as concrete blocks shall be used to support scaffolds or planks. Scaffolds shall also be tied into the structure, guyed, or out-rigged whenever their height exceeds 4 times the minimum base dimension, and/or their length exceeds 20'. Ladder jacks, lean-to, and prop scaffolds are prohibited.

Scaffold shall be equipped with a top rail made of lumber not less than 2" by 4" (or equivalent in strength), 42" high; a 21" high mid-rail made of lumber not less than 1" by 6" (or equivalent in strength); and toe-boards shall be installed on all open sides and end of scaffold platforms 10' or more above the ground or floor. Any scaffolding in excess of six feet in height requires standard railing on open sides and ends. The width of all scaffolds and walkways must be at least eighteen inches.

Scaffolds must be kept clear of ice, snow, grease, mud, or any other impediment and hazard. Equipment or material on the scaffold deck must either be removed or secured. Scaffolds and their components shall be capable of supporting without failure at least 4 times their maximum intended load. In no event shall this intended load be exceeded. Where persons are required to work or pass under the scaffolds, scaffolds shall be provided with a screen (or the equivalent) between the toe-board and the guardrail, extending along the entire opening. Personnel are not to ride rolling scaffolds. Rolling scaffolds shall only be used on smooth, level surfaces; otherwise the wheels shall be contained in wooden or iron channels which are level and stabilized.

Safe access shall be provided to the scaffold platform, specifically, a ladder with a safe means of access to the platform from the ladder. Employees working on suspending scaffolds shall wear a full-body harness with lanyards attached to an independent lifeline. No rigging from scaffold members shall be permitted unless catheads or well wheels designed for such purposes are utilized. Whenever such systems are used, the individuals performing the work shall ensure no personnel are exposed to falling material or equipment.

RIGGING EQUIPMENT

All rigging equipment shall be inspected for safety by a designated, competent General Contractor employee prior to initial use on the project and monthly thereafter. Records of each of these inspections shall be kept on-site by the contractor using the rigging equipment, and a copy forwarded to the General Contractor prior to the start of any shift. Damaged rigging equipment shall be removed from service and taken off site at time of inspection or when a safety violation is observed. A rigging plan shall be submitted to the General Contractor for review a minimum of 48 hours prior to any lift.

All cranes and derricks shall be certified as being in safe operating condition by the General Contractor, or its representative, prior to using the crane or derrick on-site. The General Contractor or its representative shall maintain this certification, and a copy sent to the subcontractor. All crane operators shall be previously qualified for that specific piece of equipment by a competent person who is capable of identifying existing and predictable hazards in the surrounding or working conditions which are unsanitary, hazardous, or dangerous to employees and who has authorization to take prompt corrective measures to eliminate them. Evidence of the operator's qualifications shall be maintained on-site and a copy sent to General Contractor.

Additional best practices relating to safety with cranes are:

- The swing radius of the crane shall be barricaded.
- Hand signals prescribed by ANSI shall be posted at the operator's station. All hand signal operators shall be properly trained.
- Manufacturer's specifications shall be strictly observed.
- Equipment shall not be operated where any of the equipment or load will come within 10' feet or electrical distribution or transmission lines.
- No person shall ride the headache ball, hook, or load being handled by the crane.
- No equipment shall be lubricated while in use.
- Rated load capacities, recommended operating speeds, special hazards warning, specific hand signal diagram, and special instructions shall be visible to the operator while at the control station.
- Employees shall not be allowed to work under the load of cranes.

FIRE PROTECTION

Fire protection and emergency equipment shall be kept free and clear from obstructions at all times and be properly located to the current work area. No gasoline, kerosene, or diesel operated equipment will be allowed inside the building at any time. Flammable fuels and materials shall be stored inside the General Contractor-provided enclosures outside the building with no more than one day's supply stored at any time. Handling shall be with proper safety containers only. Storage of combustible construction material shall not be permitted in the building during construction. Flammable debris and rubbish shall be placed in metal containers with covers and removed from the premises daily. Any plastic, tarpaulin, or other material used to construct a hut, tent, or similar protective structure, shall be flame resistant.

Fire Extinguishers

The General Contractor is responsible for providing a sufficient number of fire extinguishers of Type A, B, and C, as appropriate, and for ensuring they are maintained in an operable condition. All portable offices, tool rooms, and material storage trailers shall be equipped with at least one 20 pound dry canister fire extinguisher within 6' of every entryway to those facilities. Connections to fire hydrants, hose stations, or sprinklers are not permitted. Fire extinguishers are required for the following kinds of work activities:

- Welding, cutting, and soldering
- On-site fueling of gasoline powered equipment

PROTECTION OF THE PUBLIC

All necessary precautions shall be taken to prevent injury to the public or damage to the property of others. Precautions to be taken shall include but are not limited to the following:

- Work shall not be permitted at any area occupied by the public unless specifically permitted by the contract or in writing by the Contracting Officer. When it is necessary to maintain public use of work areas involving sidewalks, entrances to buildings, lobbies, corridors, aisles, stairways, and vehicular roadways, the General Contractor shall protect the public with appropriate guardrails, barricades, temporary fences, overhead protection, temporary partitions, shields, and adequate visibility.

- Sidewalks, entrances to buildings, lobbies, corridors, aisles, doors, or exits shall be kept clear of obstructions to permit safe entrance and exit of the public at all times.

- Appropriate warnings and instructional safety signs shall be conspicuously posted where necessary. In addition, a "signalman," furnished by the General Contractor, and the General Contractor shall control the movement of motorized equipment in areas where the public might be endangered.

- Sidewalks, sheds, canopies, catch platforms, and appropriate fences shall be provided when it is necessary to maintain public pedestrian traffic adjacent to the erection, demolition, or structure alternation of the outside walls on any structure.

- A temporary fence shall be provided around the perimeter of above ground operations adjacent to public areas. Perimeter fences shall be at least 6' high. They may be constructed of wood or metal frame and sheathing, wire mesh, or a combination of both. When the fence is adjacent to a sidewalk near a street intersection, at least the upper portion of fence shall be open wire mesh from a point not over 4' above the sidewalk and extending at least 25' in both directions from the corner of the fence or as otherwise required by local conditions.

- Guardrails shall be provided on both sides of vehicular and pedestrian bridges, ramps, runways, and platforms. Pedestrian walkways elevated above adjoining surfaces, or walkways within 6' of the top of excavated slopes or vertical banks shall be protected with guardrails. Guardrails shall be made of rigid materials capable of withstanding a force of at least 200 pounds applied in any direction, at any point in their structure. They shall be 2" by 4" dressed wood or the equivalent. Intermediate horizontal rails at mid-height and toe-boards at platform level may be 1" by 6" wood or the equivalent. Posts shall not be over 8' apart.

- Barricades meeting local requirements shall be provided where applicable to protect public users in or around your project. Barricades shall be secured against accidental displacement and shall be maintained in place except where temporary removal is necessary to perform the work. During the period a barricade is temporarily removed for the purpose of work, a "watchman," furnished by the General Contractor, shall be in place at all openings to prevent access from all unauthorized parties.

- Temporary sidewalks shall be provided when a permanent sidewalk is obstructed by the General Contractor's operation. They shall be installed in accordance with the requirements listed above.

- Warning lights shall be maintained from dusk to sunrise around temporary walkways in both public and construction areas, excavations, barricades, or obstruction in the public areas.

FALL PROTECTION

All employees shall wear a full body harness when working 6' or more above the ground when no other type of fall protection is provided. The lanyard shall be securely attached to the employee 100% of the time and shall **allow a maximum fall distance of 6'**. A full body harness shall also be worn and attached to the tie-off rail when working out of extensionable and articulating boom platforms or suspended scaffolding. Safety nets shall also be provided when work places are more than 25' above the ground or where the use of other fall protection devices are impractical.

STEEL ERECTION

Standard guardrails and toe-boards shall be installed around the open sides of permanent floors, decks, and mezzanine levels. During structural steel assembly, a safety railing (cable) of one half (1/2) inch in diameter shall be installed approximately 42" inches high, recessed into the interior of the floor and surrounding all temporary floors. A second cable shall be installed approximately 24" high.

Where the fall distance from an unprotected floor or wall opening exceeds 6', scaffolds, ladders, catch platforms, or a full body harness with lanyards attached to lifelines or other substantial objects shall be used in accordance with the procedures outlined for general fall protection. If the use of these is impractical, safety nets shall be provided.

Multi-sling lifts (Christmas treeing) are not permitted.

DISASTER PLANNING

Unless provided for elsewhere in the documents, each General Contractor shall be responsible for reasonable and prudent precautions for the protection of the personnel, equipment, materials, and installed work from weather and other acts of God.

SIGNS, SIGNALS, AND BARRICADES

Signs, signals, and barricades shall be visible at all times where a hazard exists and removed promptly when the hazard no longer exists.

HAND AND POWER TOOLS

All hand and power tools, whether purchased by the General Contractor or by the General Contractor's employee, shall be maintained in a safe condition. Under no circumstance shall the General Contractor issue nor permit the use of unsafe hand or power tools. All hand tools shall be inspected monthly by the General Contractor and anything determined to be damaged shall be removed from the site immediately. Only properly trained employees shall operate power-actuated tools. Employee certification records shall be maintained by the General Contractor on each employee using power-actuated tools on the project. All tools shall conform to OSHA and ANSI requirements.

All pneumatic power tools shall be secured to the hose or whip, or by some other positive means to prevent whipping when air pressure is applied to the line. All cords, leads, and hoses shall be kept off the ground at least 7' or whatever height is necessary to be protected from traffic. All electric hand tools shall be properly grounded and Ground Fault Interrupters (GFI's) shall be located on all electrical panels providing temporary power. All compressed air hose connection fitting(s) shall be wired prior to use.

COMPRESSED GAS CYLINDERS

Cutting, burning, and/or welding permits must be issued prior to the commencement of any such work. Upon completion, the work area will be examined by the General Contractor for sparks or embers. The permits are then signed and returned to the appropriate authority. A fire watch shall be required while cutting and burning and until all glowing embers are extinguished.

The location of cylinder storage areas must be reviewed by the Contracting Officer and must contain "No Smoking" signs, fire extinguishers, and signage clearly indicating the contents of cylinders. Compressed gas cylinders shall be secured in an upright position at all times. When transporting, moving, and storing cylinders, they shall be kept in the upright position and properly tied off. Valve protection caps shall be in place and secured. Cylinder markings will be exposed to pedestrian traffic and shall be labeled as to the nature of their contents. Cylinders shall be kept away from sparks, hot slag, and flames, or be adequately protected. Cylinders shall not be hoisted by magnets or choker slings. Valve protection caps shall not be used for hoisting cylinders.

Oxygen cylinders in storage shall be separated from fuel gas cylinders or combustible materials a minimum of 20', or by a noncombustible barrier at least 5' high having a fire-resistant rating of at least one-half hour. Back flow valves are required on all oxygen and acetylene lines. Cylinders shall not be placed where they can become part of an electrical circuit.

LADDERS

Always check with the specifications to ensure you are using an approved type of ladder prior to deploying them to the job site. The use of ladders with broken or missing rungs or steps, broken or split rails, or other defects is prohibited. Portable ladders shall be equipped with safety shoes. The use of a ladder is prohibited when wind speeds (or gusts) exceed 20 m.p.h.

Ladders shall extend no less than 36" above any landing and shall be secured to prevent displacement. All ladders shall be properly tied-off and never set up on stairways. While ascending or descending ladders, the user shall always face the ladder and nothing shall be carried that will prevent the user from holding on with both hands. A hand-line shall be used if it is necessary to raise or lower materials.

Ladders shall not be stored in the upright position when not in use.

EQUIPMENT AND MOTOR VEHICLES

All operators of construction equipment shall be previously qualified by the General Contractor with evidence of the operator's qualifications maintained on-site. All equipment shall be inspected daily by the General Contractor before use by the operator for visible and obvious hazards such as lights, seatbelts, sirens, and obstructions. Formal inspections shall also be made at regular monthly intervals for more comprehensive items like fluids, lubricants, treads, and rotations. The General Contractor shall maintain records of these inspections on-site, and copies shall be made available upon request. Defective equipment shall be repaired, when practical, or removed from service immediately. Also, fire extinguishers shall be mounted on all motorized vehicles, with the proper inspections sticker or tag attached.

All rubber-tired, self-propelled scrapers, rubber-tired front-end loaders, rubber-tired dozers, wheel-type agricultural and industrial tractors, crawler tractors, crawler-type loaders, forklifts, and motor graders shall be equipped with rollover protective structures and seat belts. Vehicles used to transport employees shall have seats firmly secured. The number of seats shall be adequate for the number of employees to be carried and all passengers shall be properly seated. Standing or riding on tailgates of moving vehicles is prohibited.

All equipment with an obstructed view to the rear shall have a reverse signal alarm audible above the surrounding noise level or must have a dedicated flagman. Vehicles with cracked and broken glass shall have the cracked or broken glass replaced before bringing vehicles on the job site. If glass is broken or damaged on-site and if damage is severe enough to cause a potential safety problem, the machine shall be stopped until such damage has been repaired. The General Contractor shall remove equipment with broken, cracked, or defective glass as required by the Owner.

Locations for storage of all fuels, lubricants, starting fluids, etc., shall be secured and barricaded. Fire extinguishers shall be provided at all such locations. Proper storage and security of these materials is subject to the review of the General Contractor.

FLOOR AND WALL OPENINGS AND STAIRWAYS

Floor and wall openings shall either be guarded by a temporary standard OSHA guardrail and toe-board, or adequately covered. When using guardrails, they shall be of sufficient strength to support a minimum of 200 pounds of pressure when applied to the midspan parallel with the floor and perpendicular to the guardrail. Covers shall be positively secured to prevent displacement and have "Danger – Do Not Remove" signs attached identifying the hazard.

Every flight of stairs having 4 or more risers shall be equipped with standard stair railings. Stairs shall not be used until risers and railings are securely installed. Where poured treads apply, the installation should be done as soon as possible after the stairs are complete. Debris and other loose material shall not be allowed on stairways or at access points to the stairway. Debris shall not be allowed to accumulate in stairwells and it is the responsibility of the General Contractor to remove any debris immediately.

CONFINED AREAS OR SPACES

All General Contractors shall develop entry procedures to be used when employees are required to enter confined areas or spaces. This procedure shall be submitted to the General Contractor for review and comments. Each entry shall be coordinated with the General Contractor prior to entry into confined areas or spaces. Such areas include storage tanks, process vessels, bins, boilers, ventilation or exhaust ducts, sewers, underground utility vaults, tunnels, pipelines, open-topped pits, basements, and temporary wood framing covered with plastic.

All employees required to enter a confined area or space shall be trained and qualified by the General Contractor as to the nature of the hazards involved, necessary precautions to be taken, and in the use of the appropriate protective emergency equipment. Before employees are permitted entry into any confined area or space, or in the event of a change in the work environment, the atmosphere within the space shall be tested to determine the oxygen level and concentrations of flammable vapors, gasses, or toxic contaminants. The General Contractor needing access to the confined area shall furnish the testing equipment and a competent person trained in the use of the testing equipment.

When welding, cutting, or heating in confined areas or space, adequate ventilation shall be provided by the General Contractor. When sufficient ventilation cannot be provided without blocking the means of access, employees shall be protected by air-line respirators and an employee shall be stationed outside the confined area to maintain communication with those working within and to aid them in the event of an emergency. All breathing air shall be provided by the General Contractor and meet all the requirements of OSHA 1926:800.

HIGHWAY WORK

All work on or adjacent to existing public roadways shall be performed in accordance with the requirements set forth in the most recent version of the *Manual on Uniform Traffic Control Devices for Streets and Highways (MUTCD)*. General Contractors should also consult local and state requirements, and in all cases, the most stringent standard shall apply where conflicts arise. Unless otherwise specified in the contract documents, the General Contractor performing such work shall be responsible for furnishing, setting up, and maintaining all necessary traffic control signage, devices, barricades, arrow boards, and flaggers.

CHAPTER 12

QUALITY CONTROL

The goal of a quality assurance system is to avoid inconsistencies where the standard of craftsmanship and resources doesn't align with specifications and drawings. The responsibility for quality construction rests with the Project Manager, QC Manager, and the Chain of Command. The Quality Control Manager is responsible for oversight on tests and inspections to ensure compliance with the plans and specs. The project superintendent will plan quality into the project and avoid discrepancies before the QC inspector performs their inspections. Discrepancies by the QC inspector represents failure in the project's QC plan.

THREE PHASES OF QUALITY CONTROL ON FEDERAL PROJECTS

If you are going to participate in construction projects with the federal government, which is our primary specialty, it is first important to be aware of the three-phase quality control approach you must follow. These guidelines have been developed by the Army Corps of Engineers and can be applied to all types of construction projects across various applications, but are non-negotiable when working with a government entity.

Initial Phase

This phase begins at the beginning of a definable feature of work. You must notify the Client and any other appropriate persons at least 24 hours in advance of the meeting to ensure their availability and ability to attend if they so wish. Required work items include:

1) Check preliminary work.
2) Check new work for compliance with contract documents.
3) Review control testing.
4) Establish level of workmanship.
5) Check for use of defective or damaged materials.
6) Check for omissions and resolve any differences of interpretation with the Client representative.
7) General check of dimensional requirements.
8) Check safety compliance.

Preparatory Phase

This phase is to be performed prior to beginning each definable feature of work. You must notify the Client and any other appropriate persons at least 24 hours in advance of the meeting to ensure their availability and ability to attend if they so wish. Required work items include (see Appendix 14-1 for the full Preparatory Phase Checklist):

1) Review contract requirements.
2) Check to assure that all materials and/or equipment are on hand and have been tested, submitted, and approved as required.

3) Check to assure that provisions have been made to provide required control testing.

4) Examine the work area to assure all preliminary work has been accomplished.

5) Review hazard analysis.

Follow-Up Phase

Perform daily checks to assure continued compliance with workmanship established at the initial phase.

1) Assurance of continuous compliance with contract drawings and specifications.

2) Daily control testing.

CUSTOMER INTERFACE

Various entities are responsible for inspection and surveillance of work. These entities approve field change directives (FCD) and respond to requests for information (RFI). The Customer will be interfacing with you to accept the work your crew has provided. It is your responsibility to prepare for such an interface and ensure that a large punch list of rework items are avoided. Try to view a project from the Customer's point of view and look at the quality of each piece and part. This is the only way to satisfactorily complete a project.

PRE-CONSTRUCTION CONFERENCES

Prior to commencing work on any project, the project manager must hold a preconstruction meeting with all parties involved (customer, General Contractor, etc.). The purpose of this meeting is to discuss the scope of the project, the construction schedule, utility requirements, the QC plan, and any other items which may affect the project. The project manager will head up this meeting and maintain minutes of the meeting to ensure the proper directives are followed. A sample pre-construction worksheet is provided in this chapter.

MATERIAL TESTING AND INSPECTION

Any material tests required by the specifications will be performed by your respective engineering department. The project manager should include these tests in his QC plan and coordinate with engineering to ensure it happens. Inspection of the materials to ensure compliance with plans and specs is the QC Manager's responsibility. These inspections must be done when materials are received and prior to installation. It is important to review key items like expiration dates, potential storage damage, and multiple other QC requirements. Whether delegated or not, the project manager must ensure material is properly inspected prior to installation.

ROLES AND RESPONSIBILITIES

Everyone involved in a construction project has very specific duties and responsibilities, and this holds true from the outset of the project all the way through quality control. It takes a complete team effort to ensure everything gets done in the most efficient and cost-effective manner and in accordance with the project plans and specifications. The quality control phase of the project is one of the last checkpoints before turning over a job to the Client, which makes it the most important for showing up professionally and fulfilling your end of the contract. Any flaws or defects are likely to be picked up by someone at some point before final payment is released, so it is best for each member of the team to know exactly what they are responsible for.

Quality Control Manager

The Quality Control Manager is responsible for creating a robust QC program for the project to verify that construction quality aligns with plan requirements and specifications. The development and implementation of the QC plan can be broken down into the following steps:

Establish Quality Measures	→	Select Construction Methods	→	Identify Required Training & Equipment	→	Ensure Personnel Awareness	→	Evaluation of Work Completed

1) **Establish Quality Measures:** The Quality Control Manager must review the plans and specs and identify the quality criteria, which must be complied with. Quality measures must be specific, giving tolerances for work performance. The language used should be "plain language" and concise. They will serve on the front end of all submittal reviews to ensure all material complies and set the standards for what will be tested, the frequency, and methods of testing.

2) **Select Construction Methods:** Effective construction techniques are vital for ensuring both safety and excellence in building projects. Construction approaches need to be established during the initial planning phase since they directly affect equipment selection, tool requirements, material choices, workforce needs, skill development, and safety protocols. The construction techniques chosen during

planning stages significantly influence the end product's quality. Standard building practices have evolved through years of repeated implementation by industry professionals. These time-tested methods typically represent the most reliable approach to achieving safe, superior results. Always surround yourself with the necessary expertise and resources to implement as many of these standards as possible. When uncertain, seek input from your team members, supervisors, and quality control personnel.

3) **Identify Required Training and Equipment:** Numerous tasks demand specific certifications or expertise. Certain ones, like welding permits or electrical connections, can only be obtained through structured education. Traditional training for various tasks may not be feasible. However, it's crucial to determine required competencies and find alternative learning methods. The primary source of instruction typically comes from on-the-job training (OJT). Select at least one skilled individual who can demonstrate the correct procedure and allocate sufficient time during the activity to instruct other team members in proper methods. Keep in mind that projects serve as learning opportunities for our workforce, so there is an emphasis on only having others who have demonstrated excellence at a task teaching others. Ensuring your team learns correct procedures and methods should be a top priority. Essential tools must also be on hand to execute the task using the chosen approach.

4) **Ensure Personnel Awareness:** In order to perform the work satisfactorily, the crew must understand what the quality measures are. Prior to commencing work on an

activity all crew members should be briefed about critical measurements, inspection items, potential problems, and each member's responsibility for quality. Use the crew briefing checklist to address these items. The Quality Control Manager is also responsible for producing meeting agendas and tracking the minutes for weekly or bi-weekly meetings with all necessary stakeholders.

5) **Evaluation of Work Completed:** Regularly scheduled QC checks are required for all projects. This report aims to record that all necessary verifications, evaluations, and examinations were conducted and confirms whether the ongoing or finished work meets or fails to meet the requirements.

DOCUMENTATION

The Quality Control Manager will maintain current records of all control activities and tests. These will include factual evidence that all required control phases and tests have been performed, including the number of tests and the results. The nature of any defects, causes for rejection, proposed remedial action, and corrective actions taken will be clearly outlined. The contractor's records will cover both conforming and defective features and will include a statement confirming all supplies and materials incorporated in the work are in full compliance with the terms of the contract. All checks and tests should be listed on the CASS. The Quality Control Manager will create and adjust the punch list throughout the pre-final and final inspection phases of the project. They will also test if all installed systems are functional and working together in harmony. Legible copies of these records on an appropriate form will be furnished to the Client daily.

The planned inspections shall be documented on the Weekly Inspection form found in Appendix 10. The site superintendent shall complete this form one week ahead. It shall be presented in the weekly coordination meetings when the construction team discusses the "one-week look ahead." The planned inspections form shall be forwarded to the Project Manager. The Project Manager will schedule designers and/or Clients to be on-site to observe inspections or witness tests. The independent testing laboratory will be informed and scheduled to perform or witness required tests. Inspections that have been made within one week shall be documented. It should be noted if the system reviewed passed inspection along with the date it was reviewed.

If a deficiency is found, the Quality Control manager shall document it on the Deficiency Action Form (see form in Appendix 11), and record the form on the deficiency-tracking log. If required, logs will be transmitted to the Client for review. When the deficiency is corrected, a follow-up inspection shall be made. If the deficiency has not been corrected, the deficiency log does not get updated. The Quality Control manager will keep the log on-site and update it weekly. The log will be available for the Clients to review on-site and, if requested, it can be transmitted.

Design Group Responsibility

After the Client accepts the cost for the project, the design team prepares and dates the "Ready for Construction Documents." All amendments issued during the pricing phase are incorporated into one set of documents. The changes are highlighted. The Ready for Construction Documents are issued to contractor's first-tier subcontractors and vendors along with their contracts. Once the construction process begins, an appropriate design team member must be designated to visit the site and assure:

- The location of the structural elements are correct: foundations, beams, and columns.
- The project is located correctly on the site prior to the installation of structural items.
- All items and devices to be located within wall cavities have been installed correctly, prior to being encased within the walls.
- All above-the-ceiling items and devices have been installed correctly prior to the ceiling being closed up.
- The start-up of the HVAC mechanical systems.

PUNCH LIST

When the project is determined to be substantially complete, the primary focus shifts to walking the job site as part of a concerted team effort with the Superintendent and Client to create a thorough punch list of outstanding items. Punch lists comments shall be organized by:

- Rooms or areas
- Exterior elevations
- Grounds
- Roof

Punch list items should be sorted by subcontractor responsibility. The Client should be present during the formal walk-through. Written reports shall be prepared for each inspection outlining the findings. To not delay progress, a verbal report shall be given to the site superintendent immediately after the inspections. A written report shall be given to the project manager and the site superintendent within one day of the findings.

SHOP DRAWING REVIEW

Shop drawings shall be reviewed to assure compliance to the contract documents with the Client's comments included. The shop drawing review for a system shall be completed within ten calendar days from receipt. Comments shall be written directly on each copy of the shop drawing–General Contractor designer comments in red and Owner comments in blue or green. To prevent separation from the shop drawing, review comments should not be typed and instead attached to the shop drawing.

Construction Administration Group

The responsibility of the Construction Administration Group toward quality assurance is to set up distribution logs for shop drawings, contracts, correspondence, contract changes, and requests for information. See Appendix 6 for samples of these logs.

Project Manager Responsibility

The project manager is responsible for ensuring that only the Ready for Construction documents are present on the job site. The project manager shall conduct a preconstruction meeting with subcontractors and vendors to review quality assurance procedures. During this meeting, subcontractors and vendors shall be informed of their roles and responsibilities, reporting structure, and level of authority.

The project manager shall also provide written notification to the commercial testing agency, outlining the scope of work for testing and inspection during the project and identifying the reporting chain of command. A sample notification is provided in Appendix 7.

SUBMITTAL REGISTER

A submittal register will be prepared listing each item of equipment and material for which submittals are required. Two completed copies will be submitted for approval within 30 days of notice to proceed. The submittal register will become the scheduling document and will be used to control submittals on this contract. This register and the progress schedules will be coordinated. Submittals covering component items that are interrelated will be coordinated and submitted concurrently.

The Submittal Register will be reviewed at least every 10 days and an updated list of all past due submittals will be submitted to the Construction Manager. Amended dates will be furnished and corrective action will be noted. A complete updated Submittal Register will be furnished upon request to the Owner or his CM Agency. The Quality Control Manager is the only person allowed to record changes and approvals. This section will detail the items that need to be recorded on the Submittal Register.

Quality Assurance

The General Contractor Project Manager along with the Project Assistant (PA) will assist with the scheduling and control of all submittals, including deviations from plans or specifications. Their assistance can prove vital in ensuring subcontractors submit the required documentation in a timely manner, but the following stages of workflow will be owned by the quality control manager. If an RFI or RFP is needed, the Quality Control Manager will coordinate with the appropriate trade. Depending on what is stipulated in the project specifications, the Quality Control Manager may be responsible for approving or declining submissions, or forwarding them to the next appropriate person in the process.

The submittal register will be coordinated with the critical path (CP) schedule. 5 copies of all approval submittals will be submitted. 3 copies of all "for information only" (FIO) level submittals will be submitted. FIO submittals will be submitted at least 2 weeks prior to procurement of contractor's certified material, equipment, etc.

The Project Manager will certify as correct and in strict conformance with contract drawings and specifications each submittal received. A General Contractor approved submittal will be used to verify that material received and used on the job are the same as approved in the contract drawings. Samples will be properly stored at the site until the Owner or Construction Manager accepts all work.

Submittals, which include proposed deviations, will have the column "variation" on the submittal form checked or indicated. The reason for the deviation will be explained and the deviation will be clearly annotated on the submittal.

The Client will return a dated and stamped copy of the approved submittal. FIO submittals will normally not be returned.

Quality Control Plan (QCP)

As noted at the beginning of the chapter and in Chapter 4, a document with the required elements of your quality control specifications or project documents will be created and logged on the submittal register for use throughout the construction process, with significant emphasis here in the quality control phase of a project.

Red-line Drawings

The Quality Control manager is responsible for maintaining a set of drawings on the project site containing any field changes marked in red. These red-lines must be updated every week. At project completion, the red-lines will be turned into the A/E. They will reflect all the approved changes and be integrated into a final set of as-builts. Included with each red line mark-ups on the drawings shall be the date the change was started along with who authorized the change. Their purpose is to serve as a permanent record for the Owner regarding all actual conditions relative to those originally designed, to note dimensional deviations not documented anywhere else, and to consolidate the *identification* of the modifications that have occurred throughout the construction period. The information is used to aid in future design, construction, and maintenance. As-Built Drawings are *not* there to repeat the detailed information of any change that is properly documented in the respective files. Immediately at the start of the project, one complete set of plans and specifications is to be sent to the job site clearly marked as As-Builts. These documents are *not* to be used for construction. They are to be properly filed and kept in good condition.

Shop Drawings

All items listed on the Submittal Register or specified in other sections of the specifications will be submitted. Each submittal will be complete and in sufficient detail to determine compliance with contract requirements.

Materials Certification

Copies of all purchase orders or subcontracts requiring receiving inspection will be given to the quality control department for receiving and record purposes. When a purchase order requires vendor certification for materials, equipment, or supplies, the certification must be verified for accuracy and compliance. Once verified, it may serve as a substitute for testing the properties specified in the certification. Copies of all certifications received will be maintained in the quality control folder and will be available to Clients upon request or submitted to them as provided in the contract specifications.

RECEIVING AND WAREHOUSING

The Quality Control Manager, or other contractor personnel, will perform inspections of all permanent construction materials received. Visual inspection must be made for:

- Identification
- Damage
- Completeness
- Evidence of compliance with approvals
- Proper documentation

The outcomes of the receiving inspection will be documented using the appropriate report form and provided to the Client for review.

OFFSITE CONTROL

Offsite fabricators and suppliers will be surveyed as necessary to ensure compliance with all contract drawings and specifications, and to verify the delivery of quality products. The findings of each survey will be documented on the appropriate form and made available to the Client. Any identified deficiencies will be communicated to the fabricator or supplier, who will be required to submit a report detailing the corrective actions taken. The contractor will inform the Client of offsite surveys.

It will be determined during the design phase whether or not a design professional will need to travel to a point of manufacturing to inspect a product prior to shipping to the job site. If so, the inspection results shall be followed up with a deficiency letter, which is to be given to the project manager. The Project Manager will transmit the letter to the manufacturer. The letter will be completed within one day of completing the inspection.

CALIBRATION OF EQUIPMENT

All contractor-furnished measuring and test equipment shall be calibrated and maintained to traceable Client standards. Records of these calibration certifications will be maintained by the Quality Control Department and made available to the Client upon request. Each instrument will be clearly and permanently marked with a unique identification number. Operation of the equipment will be limited to authorized personnel or those under their direct supervision. Every piece of equipment will be inspected for accuracy at intervals recommended by the manufacturer for calibration. A certified laboratory will conduct required calibration of measuring and test equipment. Any measuring and test equipment dropped, damaged, or believed to be inaccurate will be removed from services and recalibrated.

FINAL INSPECTION AND TEST

Prior to final inspection or start of tests, all systems being inspected or tested shall be completed and accepted by the Quality Control Manager. After this acceptance, the final inspection and test may proceed in accordance with the following steps:

1) Verify the test personnel have a working knowledge of the special characteristics of the instruments being used.

2) Note the particular inspection or test requirements and criteria for successful completion of the required inspection or test.

3) Upon satisfactory verification of these requirements, the test may proceed. Each reading will be verified and documented by the Quality Control Manager. The Quality Control Department will perform all functional validations or tests unless otherwise noted. The Quality Control Department will perform no functional test unless otherwise noted. No functional test will be accepted without properly authorized and approved test procedures.

4) The general requirement of final acceptance will include, but not be limited to, the following:

a. General appearance
b. Workmanship
c. Cleanliness of areas and equipment
d. Identification of equipment
e. Painting
f. Removal of unused material and temporary facilities
g. Condition of job files and completion of paperwork

REVISION POLICY

Activities, programs, and procedures not covered in the Quality Control Plan (QCP), proposals, or additions to these standards, shall be discussed at meetings held for that purpose at such times and places the Quality Control Manager may select, and shall take such action to request acceptance from the Client to incorporate such revisions as deemed necessary. A record shall be kept of such meetings and interested parties present, together with the subject matter reviewed. Such meetings shall be held as required by changes in the contract specifications for the purpose of reviewing the QC plan, to entertain revisions, additions, or deletions. Accepted revisions shall be incorporated in the plan as first revision, second revision, etc., and a revised index page shall be included.

CHAPTER 13
PHYSICAL & CYBERSECURITY

Security is always a concern when it comes to construction projects for many reasons. There are the obvious ones of safety and theft, but when working with federal government contracts the stakes are increased drastically and it is the responsibility of everyone involved to make both physical and cyber security of paramount importance.

PHYSICAL

Government projects are usually sensitive in nature, leaving a greater risk of physical security breaches that could compromise the job site. One of the first precautions a General Contractor should take prior to any work beginning is establishing a perimeter with fencing to prevent people from gaining access. The flow of people and personnel into and out of the site should be controlled using designated access points staffed by physical security guards with an established protocol for allowing entry. This can include swiping access badges or having a manual sign-in process where all entrants are cross-referenced against a list of authorized parties.

Within the main physical perimeter, conditions may require establishing a secondary perimeter with additional access points and security checkpoints for controlled materials. This will help ensure only personnel with the proper level of security clearance are able to access items that may be deemed sensitive or of high value. Those with clearance should be tasked with inspecting these materials on a routine basis to prevent damage, theft, or espionage. Another general best practice to prevent situations like this is to have a designated party responsible for walking the site and remaining vigilant.

The last element of physical security to take into account is the monitoring of the actual work being put in place. Depending on the level of clearance needed to work on certain projects or in designated areas, it might be feasible to hire contractors with those credentials. When that happens, someone must be assigned to physically watch them work at all times in those designated areas. When personnel leave for lunch or breaks, someone with proper clearance must escort them without leaving the remaining workers unattended. On most job sites, there are three types of physical security personnel you are likely to encounter or need.

Unarmed Security Guard

A security officer is often the very first and very last impression an employee, guest, tenant, or visitor will have of your Company, so it is critical to hire people that are not only capable of doing the job but who also have a professional and friendly image as people both arrive and depart the facilities. Upwards of 90% of an officer's day-to-day job involves delivering world class Customer service with very little time ever spent dealing with any security threats.

A good security guard will be capable of controlling access points to Company property to prevent unauthorized entry and illegal removal of assets. Making positive eye contact with each person and acknowledging them as they enter the property, issuing a polite verbal greeting and assisting with questions, directions, etc., not only presents a professional image but it also serves as a deterrent to those who might have malicious intentions because they know they have been seen.

When not stationed at the entry points of a job site or facility, guards should conduct foot patrols throughout the interior and around the perimeter of the property. The purpose of these patrols is to prevent unauthorized access through alternate entry points and to deter or identify unauthorized or illegal activities such as security breaches, theft, pilferage, vandalism, or misconduct that could harm the Company, its tenants, or its employees. Additionally, these patrols help detect and prevent serious maintenance issues, including water leaks, equipment malfunctions, and similar concerns. The detection of any unauthorized persons found on Company property, whether it be in person or via the monitoring of closed circuit television systems should be addressed by following established protocols as soon as possible.

Guards should also operate, monitor, and maintain all other safety and security systems, remaining vigilant for suspicious activity, individuals, and vehicles, as well as for safety hazards or incidents, and take appropriate action to correct related problems. This could range from issuing parking citations to illegally parked vehicles, to calling maintenance when needed for urgent repair issues, or even subduing individuals when the situation warrants such physical intervention.

Cleared Escort

Much like security guards, a foundational responsibility for a cleared escort is to ensure that only persons with a valid need and acceptable authorization are permitted to enter the facility. They will also escort and monitor operational vendors at all times while on-site, maintaining line-of-sight observation at all times. Any vendors or other authorized visitors will be recorded on the visitor access logs consistent with the requirements of each individual job site. In working with those coming on-site, the cleared escort will place an emphasis on preventing the unauthorized disclosure of sensitive information.

From a maintenance standpoint, cleared escorts will escalate equipment issues found by vendors immediately to the on-site security manager. They are also responsible for properly inspecting secured areas prior to escorting uncleared personnel into the facility to remove all instances of information or technology that could be considered top secret or confidential in nature. Because of these sensitive duties, almost all job sites will require a cleared escort to hold an active top secret or sensitive compartmented information (SCI) clearance level.

Access Control Specialist

The access control specialist will monitor and control the access of employees and visitors in and out of restricted areas by ensuring people have proper access. This will include ensuring prohibited and restricted items do not find their way in or out of the building. Anyone performing this role must diligently monitor any restricted areas they are assigned to by remaining alert and aware of their surroundings at all time. This must be balanced by also providing excellent Customer service to all employees and visitors. To do this, an access control specialist should be able to exhibit excellent communication in every interaction, ensure attention to detail, demonstrate the ability to work independently during their scheduled shift(s), and adapt to change within high-volume, fast-paced, and slow-paced environments.

CYBERSECURITY

Cybersecurity has become one of the most significant areas of concern for businesses and individuals in recent years. The more we become reliant on computers and cloud storage for our diverse needs, the more opportunity we create for nefarious actors to find ways to steal, ransom, or otherwise compromise that which we hold so dearly. When dealing with federal government contracts, no matter the level of security clearance required, the need to stay vigilant and protect sensitive information becomes even more important.

One of the best and most sophisticated ways to achieve this is by adhering to the Cybersecurity Maturity Model Certification (CMMC), which outlines procedures for better controlling sensitive information and holding anyone with access to unclassified information more accountable. A Level I certification focuses on security around passwords and file sharing. The Level II and Level III certifications are more stringent and require an outside source to come in and audit the systems and procedures in place for compliance.

Then there are the Controlled Unclassified Information (CUI) procedures to adhere to. This exists because not everything done under the federal government umbrella will be considered fully classified. Projects such as embassies and military bases where weapons or active personnel will be stored or stationed would be highly classified for obvious reasons, making it difficult for anyone without proper credentials to access them. However, other projects like vehicle maintenance facilities or general material storage buildings would not be as stringently protected. So, it is the responsibility of the General Contractor to limit who they allow to have access to these drawings and specifications during and after the project.

Given the numerous considerations around cybersecurity and the inherent sensitivity of working on projects for the federal government, this is not an area to be taken lightly. If you do not have a Chief Information Officer, Chief Technology Officer, or other specialists on your team, make sure to familiarize yourself with the threats and remedies for:

- Access to sensitive data and related concerns
- Compromised supply chain
- Insider threats
- Control system threats (Wi-Fi, SCADA, Alarms, VPNs, etc.)
- Ransomware attacks
- Remote work vulnerabilities

It is important to take all threats seriously on a federal construction project. If you see suspicious activity or believe your data has been compromised, contact the authorities immediately.

CHAPTER 14
PROJECT CLOSEOUT

This section will guide the project manager through the final operational and administrative tasks at project completion. It will also help you prepare for the final inspection. Considering how much time and money has been invested into getting a project to the finish line, this is not the time to miss a crucial detail.

Tool, Equipment, and Material Turn-in

The project manager must ensure the job site is clean of all tools and excess material, which should be delivered to the Client or subcontractor it belongs to. Equipment is to be properly cleaned, inventoried, and turned over to the Client or subcontractor of record. Tools and tool kits must be inventoried and signed as received from the Customer. It is imperative a project close-out report be complete with the Customer signature indicating it has been received.

AS-BUILT DRAWINGS

Maintaining Redlines During Construction

The Construction Manager General Contractor must maintain current drawings throughout the project, coordinating with the engineering staff to ensure changes are synchronized across all disciplines. At the job site, the Superintendent or Quality Control Manager shall keep an up-to-date redline markup set of drawings showing the "as-built" condition.

Each redline entry must:

- Be marked in red pencil or pen.
- Be dated.
- Reference the appropriate authority (e.g., change order file number, job meeting reference, or structural modification authorization form).
- "Clouded" for ease of reference or when additional detail on the change is warranted.

It is not necessary to fully redraft details of changes; instead, the affected area should be clouded and cross-referenced to the appropriate documentation. Copies of sketches or authorizations may be attached to the set when space allows.

Immediately at the start of the project, one complete set of plans, specifications, and addenda shall be designated as the official as-built set. These are not to be used for construction but serve as the permanent record of deviations from the original design. Even when not explicitly required by the contract, As-Built Drawings should be maintained as a best practice to document dimensional deviations, underground conditions, clarifications, and field accommodations.

Subcontractors are required to update their trade-specific as-built information on the Company field set and As-Built Drawings at least weekly. The Project Manager must verify compliance monthly as a condition of payment. At a minimum, the following trades must be included:

- Concrete
- Structural steel
- Plumbing
- HVAC
- Fire Protection
- Electrical
- Controls
- Communications

Final As-Built Deliverables at Closeout

At project closeout, the Project Manager shall submit one complete set of redline drawings to the Owner. These will be converted into the official As-Built Drawings, which serve as a permanent record of how the project was constructed and provide critical information for future design, construction, and maintenance.

Electronic files must:

- Be organized by building system or chronological sequence.
- Undergo quality verification before submission.
- Be transmitted via external storage device or secure cloud platform.

- Be cross-checked against the drawing index and electronic file list to ensure accuracy and completeness.
- Follow version control procedures exercised by each discipline to maintain integrity of design files.

Contract documents may also require Geographic Information System (GIS) drawings to be submitted. GIS Drawings displays pertinent data for project(s) in a database. Precise points of interest for your project are taken by a surveyor and then loaded into an electronic file to be incorporated into a database. This is a critical step of the project as it provides an accurate source of truth for current and future stakeholders. GIS Drawings assist in future planning and construction, maintenance, and emergency response procedures. Examples of items on your project that may be required for your GIS Drawings file would be utility locations (manholes, transformers, valves, electrical panels, etc.), building elevations, building corners and/or highly desired surveyed shots around the building, emergency service connections, hardscape elevations, and property lines.

In addition to As-Built Drawings, all Operation and Maintenance (O&M) manuals, test reports, and training certificates shall be compiled and transmitted to the Owner. These must be:

- Organized by building system.
- Delivered no later than substantial completion.
- Verified by the Project Manager to ensure all files are present and accessible.

Some projects may require a Warranty Management Plan or a Project Closeout Plan. These plans will detail proper warranty and closeout procedures. This is imperative to complete as often times projects can drag out during closeout. Having a plan in place ensures that all the requirements for the end user are met and they are left with the proper training, closeout, and warranty information needed for successful operation of their project.

Some contracts may require the final as-built record to be transferred to specific media (e.g., mylar) or include certified as-builts from particular trades such as Fire Protection. Review contract documents carefully to confirm and comply with all Owner-specific requirements.

Final As-Built Drawings and associated deliverables should be hand-delivered to the Owner whenever possible, with receipt acknowledged by an authorized representative.

Acceptance

At the completion of your project, use the project close-out report to make sure your project is ready, then arrange for a preliminary acceptance inspection with the Customer. During this inspection a punch list is created. These items must be rectified in order to go to the next step, which is the final acceptance inspection.

In most cases, the project will not be turned over to the Customer until all punch list items have been completed. When all the punch list items have been completed, the project manager must schedule a final acceptance inspection to occur. There should be no punch list items at this inspection. If there are no discrepancies, beneficial occupancy is established upon completion of the final inspection and the one-year warranty takes effect. The close-out report should be completed in its entirety and the Customer should be notified the project is officially complete. At this time the project files should be closed and retained for future reference.

OWNER'S DOCUMENTATION REQUIREMENTS POST CONSTRUCTION PHASE

By the time we reach this stage of the project close out, we are effectively putting together all of our information to document exactly how the project went. A primary component of this will be how well we adhered to the schedule or deviated from it. Then we want a documented record of how accurately the finished appearance reflects what was called for in the contract drawings and specifications. Lastly, this is where we want to compile warranties, manuals, and everything else the Owner/client could conceivably need to take control of the completed structure without relying on the contractor's assistance going forward.

Incorporation of "Redlines Drawings" into As-Built Drawings

The quality of electronic files will be verified prior to the submittal of corrected as-built files. Version control will be exercised by each discipline, with final design files being cross-checked against the drawing index and the electronic list of file names and PDFs in order to ensure accuracy and completeness. Actual files used to plot the final submittal will be moved to a new directory to segregate the submittal files from all others. A designated representative shall be responsible for limiting access to old files and ensuring that only the most current set of submittal files remains readily available for use. After transferring to external storage options, the General

Contractor will verify they are readable and downloadable. Several test plots will be performed with the electronic version to verify size and plot quality. All work will be performed using Computer-Aided Design (CAD).

Required Submission of Hardcopy Records

Items the General Contractor is required to transmit to the Client, such as operation and maintenance manuals, test reports, certificates of training, etc., will be organized by building systems. The records will be transmitted to the Client no later than when the project is substantially completed.

Required Submission of Electronic Records

The electronic files the General Contractor is required to send will be transmitted to the Client via the pre-agreed upon electronic means no later than when the project is substantially completed. The files will be organized by either building system or chronological, whichever is applicable. After transferring or uploading, all files will be physically verified by the Project Manager and CADD support group to confirm all files are present and readable/downloadable.

Warranty

For up to 1 year following the performance of the work, the General Contractor (GC) shall re-perform, at its own expense, all services within the original scope of work, which fail to conform to the foregoing standard of care, provided the Owner notifies the General Contractor in writing within 30 days after discovery of such nonconformity. See example of a Warranty Work Order in Appendix 13-1. Here is the process warranties undergo:

Warranty Process Flowchart

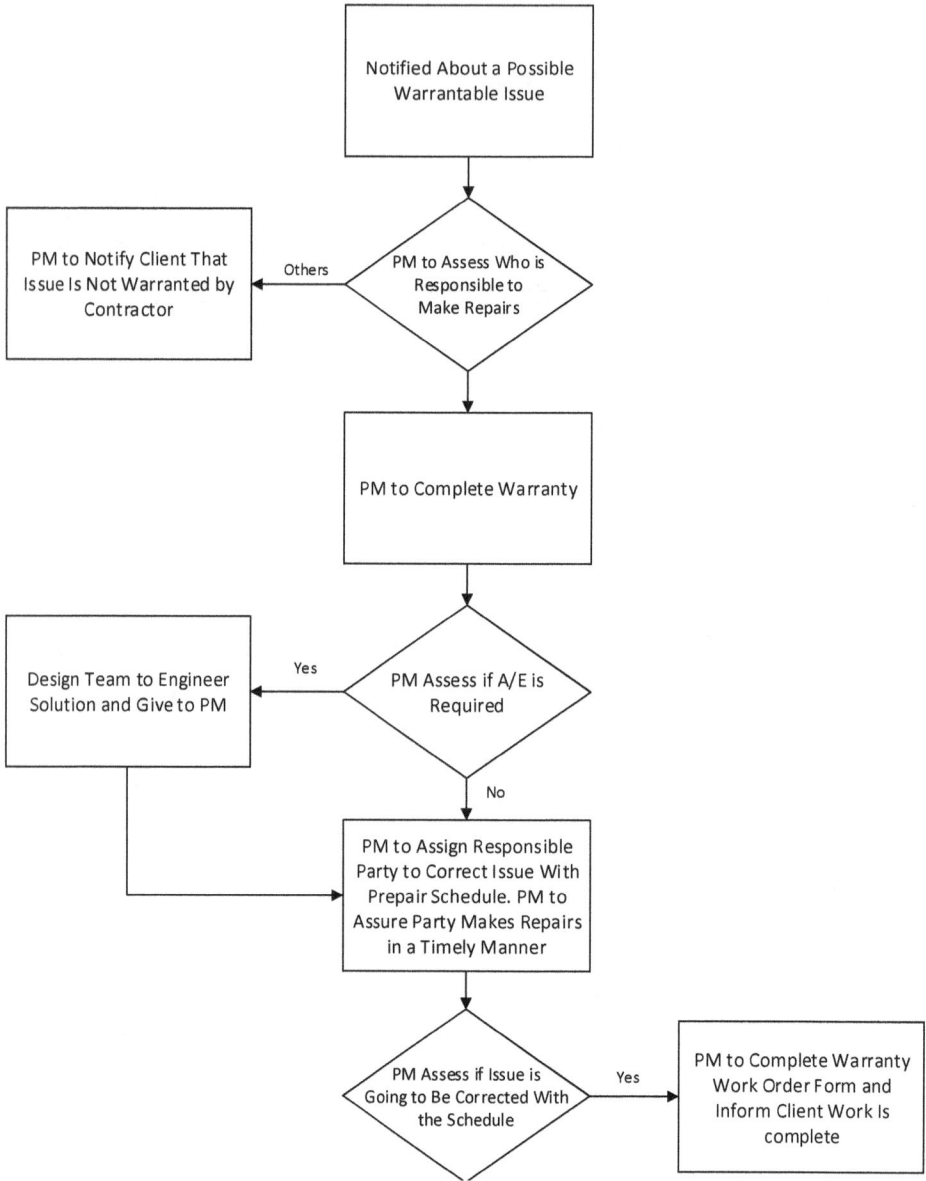

```
                    ┌─────────────────────┐
                    │ Notified About a    │
                    │ Possible            │
                    │ Warrantable Issue   │
                    └─────────┬───────────┘
                              │
                              ▼
┌──────────────────────┐   ◇─────────────◇
│ PM to Notify Client  │   │ PM to Assess │
│ That Issue Is Not    │◄──┤ Who is       │
│ Warranted by         │Others│ Responsible│
│ Contractor           │   │ to Make      │
└──────────────────────┘   │ Repairs      │
                            ◇──────┬──────◇
                                   │
                                   ▼
                    ┌─────────────────────┐
                    │ PM to Complete      │
                    │ Warranty            │
                    └─────────┬───────────┘
                              │
                              ▼
┌──────────────────────┐   ◇─────────────◇
│ Design Team to       │Yes│ PM Assess if │
│ Engineer Solution    │◄──┤ A/E is       │
│ and Give to PM       │   │ Required     │
└──────────┬───────────┘   ◇──────┬──────◇
           │                   No  │
           │                       ▼
           │         ┌─────────────────────┐
           │         │ PM to Assign        │
           └────────►│ Responsible Party   │
                     │ to Correct Issue    │
                     │ With Prepair        │
                     │ Schedule. PM to     │
                     │ Assure Party Makes  │
                     │ Repairs in a Timely │
                     │ Manner              │
                     └─────────┬───────────┘
                               │
                               ▼
                     ◇─────────────◇    ┌──────────────────┐
                     │ PM Assess if │Yes │ PM to Complete   │
                     │ Issue is     ├───►│ Warranty Work    │
                     │ Going to Be  │    │ Order Form and   │
                     │ Corrected    │    │ Inform Client    │
                     │ With the     │    │ Work Is complete │
                     │ Schedule     │    └──────────────────┘
                     ◇─────────────◇
```

Warranty Work Order

Project Superintendent: Site Administrator:

PROJECT: Request By: Date WO Requested:

PROBLEM:

General Contractor:

Date & Time On Site: Client Notified?   Yes ☐   No ☐

Who was Present: Name of individual notified:

_____

WORK ACCOMPLISHED:

DATE/TIME COMPLETED: _____ CONTRACTOR SIGNATURE: _____

NOTES TO CONSTRUCTION/MAINTENANCE PERSONNEL:

Design & Estimating Flowchart

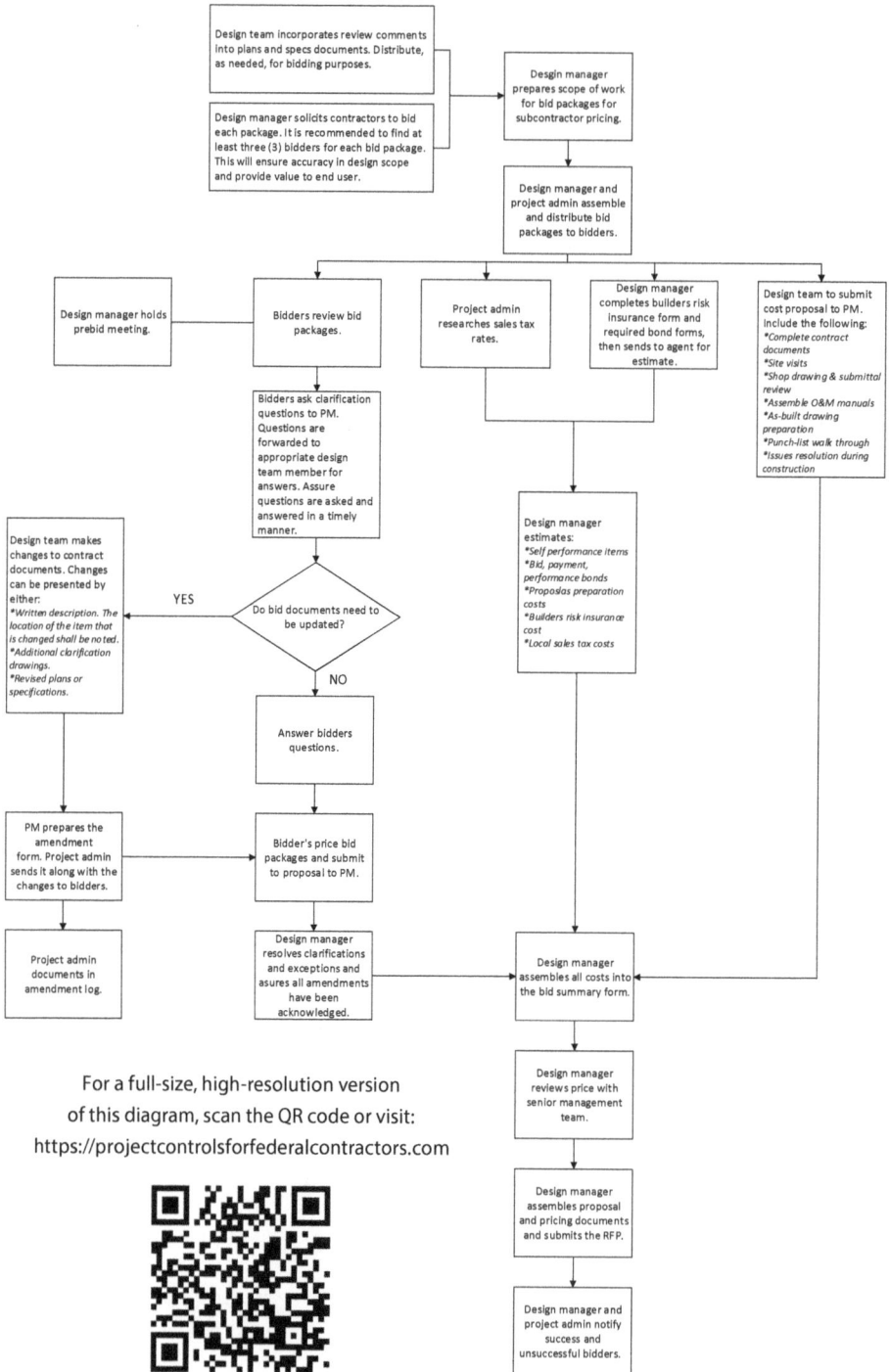

Design team incorporates review comments into plans and specs documents. Distribute, as needed, for bidding purposes.

Design manager solicits contractors to bid each package. It is recommended to find at least three (3) bidders for each bid package. This will ensure accuracy in design scope and provide value to end user.

Design manager prepares scope of work for bid packages for subcontractor pricing.

Design manager and project admin assemble and distribute bid packages to bidders.

Design manager holds prebid meeting.

Bidders review bid packages.

Project admin researches sales tax rates.

Design manager completes builders risk insurance form and required bond forms, then sends to agent for estimate.

Design team to submit cost proposal to PM. Include the following:
*Complete contract documents
*Site visits
*Shop drawing & submittal review
*Assemble O&M manuals
*As-built drawing preparation
*Punch-list walk through
*Issues resolution during construction

Bidders ask clarification questions to PM. Questions are forwarded to appropriate design team member for answers. Assure questions are asked and answered in a timely manner.

Design team makes changes to contract documents. Changes can be presented by either:
*Written description. The location of the item that is changed shall be noted.
*Additional clarification drawings.
*Revised plans or specifications.

YES

Do bid documents need to be updated?

NO

Design manager estimates:
*Self performance items
*Bid, payment, performance bonds
*Proposias preparation costs
*Builders risk insurance cost
*Local sales tax costs

Answer bidders questions.

PM prepares the amendment form. Project admin sends it along with the changes to bidders.

Bidder's price bid packages and submit to proposal to PM.

Project admin documents in amendment log.

Design manager resolves clarifications and exceptions and assures all amendments have been acknowledged.

Design manager assembles all costs into the bid summary form.

For a full-size, high-resolution version of this diagram, scan the QR code or visit: https://projectcontrolsforfederalcontractors.com

Design manager reviews price with senior management team.

Design manager assembles proposal and pricing documents and submits the RFP.

Design manager and project admin notify success and unsuccessful bidders.

CHAPTER 15

BRINGING IT ALL TOGETHER

Federal construction projects demand a higher level of planning, compliance, and documentation than almost any other sector of the industry. Throughout this book, we have explored the full life cycle of a project from bidding and estimating through environmental considerations, scheduling, material and equipment management, execution, monitoring, change management, safety, quality control, and ultimately project closeout.

Every topic covered shares a single common thread: strategic project controls. Strong project controls are not simply paperwork requirements; they are the tools that enable Project Managers, Superintendents, Quality Control Managers, and the entire project team to deliver projects safely, on time, within budget, and to the exacting quality standards required by federal clients.

As you implement the processes and principles in this guide, keep the following takeaways in mind:

- **Plan ahead and stay proactive.** Federal construction projects reward foresight and penalize reactivity. Use schedules, submittals, and resource plans as living documents, not static checklists. These items can be your project's best friend or their worst enemy.

- **Document everything.** Well-organized submittals, safety documents, redline drawings, RFIs, and change orders are not only contractual obligations but also essential protections for your team and your company. It may seem burdensome, but proper documentation will help your project run much smoother.

- **Prioritize safety and quality first.** When safety and quality are consistently managed, schedules and budgets naturally follow. Always remember that everyone has someone that depends on them. We want everyone working on our projects to make it home to their families every day.

- **Communicate across disciplines.** Project controls only work when information flows freely between design, management, subcontractors, and the Owner's representatives. Set communication expectations early and don't waiver.

- **Embrace continuous improvement.** Federal standards evolve. So do best practices in technology, sustainability, and cybersecurity. Regularly review your processes and update your tools to remain competitive.

Finally, remember that project controls for federal construction is more than compliance — it's leadership. When you apply the principles in this book, you're not just checking boxes; you're creating a culture of accountability, professionalism, and excellence. The effort you invest in planning, monitoring, and controlling each aspect of your project will pay dividends in fewer disputes, stronger client relationships, and safer, higher-quality work.

Use this book as a reference and a framework, but adapt it to the unique conditions of each project. Federal construction is complex and challenging, but it offers opportunities to develop skills, deliver lasting infrastructure, and build a reputation for reliability and expertise. By mastering project controls, you're equipping yourself and your team to meet those challenges and exceed expectations.

APPENDICES

PROJECT PACKAGE

SECTION 1: General Information –
Correspondence – Financial

1A	Contract
	RFP
	Estimating Worksheet
	Schedule of Values
1B	Project Scope Worksheet
	Project Organization
	Preconstruction Checklist
	Construction Meeting/Teleconference Notes
	Hardcopy Correspondence
	Electronic Correspondence

SECTION 2: Activities – Network

2A	Level II SITREP
	Project Schedule
2B	Construction Activity Summary Sheets
	Daily Production Reports
	General Contractor Level II Input

SECTION 3: Resources

3A	Long Lead Item Tracking Worksheet
	Special Tool Tracking Worksheet
	Special Equipment Tracking Worksheet
3B	Warranty
	Rework

SECTION 4: QC – Safety – Change Orders

4A	Quality Control Information
	Safety Information
	Environmental Information
4B	Change Order Submittal Log
	Change Orders
	Request For Information (RFI) Submittal Log
	Request For Information
	Field Adjustment Request (FAR) Submittal Log
	Field Adjustment Request (FAR)
	Design Change Directives (DCD)

SECTION 5: Plans - Specifications

5A	Project Plans
	As-Built Drawing
5B	Project Specifications

BUILDERS' RISK QUESTIONNAIRE

Named insured:	Horizon Builders, Inc.
Contact name:	John Davis
Contact phone number:	(948) 456-7890
Date quotation needed:	May 31, 2025
Anticipated project start date:	July 01, 2025
Anticipated project completion date:	December 31, 2027
Location of project:	Fort Valor Military Base, VA
Occupancy of buildings, when complete:	2,500
Client for whom project is being constructed:	US Army, Fort Valor
Is the cost of Builders' Risk to be a pass through?	No
Should subcontractors' and subcontractors' interests be included in Builders' Risk?	No
Does mortgagee's interest need to be included in Builders' Risk?	Yes
If so, please supply the name and mailing address for the issuance of a certificate of insurance:	Horizon Builders Inc. 1234 Horizon Blvd, Suite 200 Richmond, VA 23219
Completed value of building(s): *(If equipment is being provided as part of the scope of work, please declare building values separate from Equipment values, by building).:*	$180 million

If more than one building is involved, please list each building, including its intended occupancy and completed value:	**Building 1: Secure Operations Center** • Intended Occupancy: 1,500 personnel • Completed Value: $120 million **Building 2: Administrative & Support Facility** • Intended Occupancy: 1,000 personnel • Completed Value: $60 million
Is coverage needed for soft costs?	Yes
If so, please state those values separately:	$20 million in soft cost coverage
Is coverage needed for building materials in transit?	Yes
If so, please advise of the amount of coverage needed:	$5 million
Is coverage needed for off-site storage of building materials?	Yes
If so, please advise of the amount of coverage needed:	$3 million
Is coverage needed for the testing of equipment, including HVAC?	No
Construction materials: (*i.e., Metal, Cinder Block, Fire Resistive, etc.*)	Steel, concrete, glass, and fire-resistive materials
Roof construction?	Built-up roofing system
Distance from the nearest body of water:	15 miles
Wind protection? (*If in hurricane zone*)	Not applicable – Fort Valor Military Base is not located in a hurricane zone
Distance from nearest fire hydrant? (*If unprotected, distance from nearest water tower, body of water, etc.*)	Approximately 50 feet from the secure facility entrance

Will the building(s) include automatic sprinklers?	Yes

Distance from, and occupancy of, nearest existing building?
    Front?
    Back?
    Left side?
    Right side?
	• **Front:** Nearest existing secure facility is approximately 100 feet away
	• **Back/Left/Right Sides:** Open secure areas with controlled access; occupancy details available in site security documentation

Is this building being tied into the existing structure?	No
If so, are we responsible for damage to the structure adjacent to the building in the course of construction? *(Please refer to proposed contract language; if so additional information will be required)?*	N/A

APPENDIX 2-2
PERFORMANCE/PAYMENT BOND REQUEST FORM

Special Bond Forms	**Delivery Required**
	Fed. Ex. X
	Will Call
	Messenger
	Other

Contract Price	$180 million
Contract Date	May 15, 2025
General Contractor	Horizon Builders, Inc.
Owner/Obligee	US Army, Fort Valor
Address	US Army, Fort Valor
	250 Command Center Blvd, Suite 200
	Fort Valor, VA 22301
Title of Job	Fort Valor Command and Control Center
Brief Description of Work	Construction of a secure command and control facility including a secure operations center, administrative offices, and support infrastructure on Fort Valor Military Base.
Contract No.	FVC-2025-001
Location of Job	For Valor Military Base, VA
Completion Time	30 months
Liquidated Damages	As per contract specifications
Performance Bond	100%
Payment Bond	100%
Maintenance Bond	10% for 2 yr(s)
Architect/Engineer	Metro A&E Solutions, Inc.
Address	300 E. Wacker Dr. Suite 500
	Chicago, IL 60601
Bid Results	Awarded to Horizon Builders, Inc.
Remarks	None

APPENDIX 2-3

CONSTRUCTION ACTIVITY SUMMARY SHEET (CASS)

Project #:	FVC-2025-001	Title:	Fort Liberty Command and Control Center
Prepared By:	Jessica Miller [PM]	Checked By:	David Nguyen [SUP]
Activity #:	001	Activity Title:	Mobilization & Site Set-up

Description of Work
The contractor will mobilize all necessary equipment, materials, and personnel to Fort Liberty Military Base to initiate the construction of the Command and Control Center. This phase includes establishing a secure construction site with enhanced security protocols required on a federal installation. Tasks include setting up temporary facilities (site office, storage areas, and staging zones), installing perimeter fencing, access control systems, and surveillance equipment, and coordinating utility connections while ensuring compliance with federal safety and security standards. An initial safety briefing and orientation will be conducted to ensure all personnel are fully aware of the heightened security and operational protocols.

Estimated Duration:	30 Calendar Days
Actual Duration:	TBD

Assumptions (Time, Material, Environmental Conditions, Manpower)
<ul><li>Favorable weather and security conditions during mobilization.</li><li>Timely delivery of materials and equipment to the secure site.</li><li>Adequate availability of skilled labor, with personnel cleared for work on a military installation.</li><li>All necessary permits, security clearances, and approvals have been obtained.</li><li>Temporary power and utility services will be provided by the Client until permanent connections are established.</li></ul>

Safety Hazards

Identified Item	Reference #	Preventative Measure
Restricted access areas	SH-001	Establish secure perimeters with access control systems and issue clearance badges to authorized personnel.
Heavy equipment operations	SH-002	Enforce strict operational protocols, ensure certified operators, and maintain continuous supervision.
Slips/trips on uneven terrain	SH-003	Maintain clear work zones, perform regular site inspections, and promptly address any hazards.
Unauthorized access	SH-004	Implement rigorous security measures including fencing, surveillance, and security personnel presence.

Quality Control Requirements

Identified Item	Reference #	Remarks
Material Inspection	QC-001	Verify that all delivered materials conform to project specifications before use.
Temporary Facility Setup	QC-002	Inspect all temporary installations for safety compliance and proper setup.
Utility Service Setup	QC-003	Confirm that temporary utilities (water, power, communications) are fully operational.

Environmental Hazards

Identified Item	Reference #	Preventative Measure
Dust and Debris	ENV-001	Utilize water suppression techniques and install dust barriers.
Soil Erosion	ENV-002	Implement erosion control measures and install sediment barriers.
Noise Pollution	ENV-003	Limit operations during early morning/late evening hours and use noise-dampening equipment.

Equipment/Tool Resources

- Mobile cranes, forklifts, and earthmoving equipment
- Temporary security equipment (surveillance cameras, access control systems)
- Site office trailers and storage containers
- Personal protective equipment (PPE) including hard hats, high-visibility vests, non-slip footwear, and security gear
- Temporary fencing, barriers, and signage
- Utility coordination tools and communication devices

Comments

- All mobilization and security setup activities must comply with federal safety and security standards.
- Coordination with military security personnel is essential to ensure all access and clearance requirements are met.
- Any deviations from the plan will be documented and reviewed during daily safety briefings.
- Adjustments to the schedule may be required based on material delivery timelines and site-specific security protocols.

APPENDIX 2-4
PROPOSAL REVIEW FORM

Proposal Review From – BD03	Proposal No.: FVC-2025-001-PR
Date: 05/25/2025	Proposal Due Date: 05/31/2025

To:
Jane Smith, Client Representative

Client:
US Army, Fort Valor

Location:
Ft. Valor, VA

Project Scope

Development of a state-of-the-art secure Command and Control Center on Fort Valor Military Base, VA. This project comprises two primary facilities: a Secure Operations Center and an Administrative & Support Facility. The facility integrates advanced security systems, secure communications, and robust support infrastructure to meet stringent federal and military standards. Designed to support 2,500 personnel at peak operational capacity, the completed facility is valued at $180 million and will serve as a critical asset for military operations.

Business Development Lead:
Michael Green
TIC ($MM): 180
Contract Value ($): $180,000,000
Contingency ($): $10,000,000

Proposal/Project Manager:
Jessica Miller
Competitors: ABC Construction Group; XYZ Commercial Developers
Net Contribution ($): $20,000,000

Incentives: Performance bonuses for early completion and cost savings incentives as defined in the contract.

Commercial Risks

- Strict base security protocols and clearance requirements may impact site logistics

- Potential material cost escalation in a competitive federal contracting environment
- Competitive bidding environment with tight margins

Contractual Risks
- Strict milestone penalties and liquidated damages provisions
- Uncertainty in soft cost recovery
- Risk of change orders due to unforeseen secure site conditions

General Contractor Differentiators
- Proven track record on high-security federal projects and military installations
- Robust project management and quality control processes tailored for secure environments
- Innovative construction techniques and efficient scheduling optimized for federal projects
- Extensive expertise in navigating military base protocols and risk mitigation strategies

Comments: The proposal is highly competitive and presents a strong value proposition for the secure Command and Control Center project. Our project team has meticulously identified key risks and developed comprehensive contingency plans. With our integrated approach and proven experience in similar secure facility projects, we are well-positioned to win this bid and deliver the project successfully.

Required Approval(s):
- ☐ CEO: Jane Doe
- ☐ COO: Mark Johnson
- ☐ VP, Operations: Karen Lee
- ☐ Business Development Manager: Michael Green
- ☐ Business Controller: Patricia Brown

SAMPLE EXECUTIVE SUMMARY SECTION

FORT VALOR COMMAND AND CONTROL CENTER

Executive Summary

Revision 1 – Data as submitted to US Army, Fort Valor 06/25/2025

Scope

Provide comprehensive Project Management, Engineering, Procurement, Construction, and Commissioning services for a state-of-the-art secure Command and Control Center located on Fort Valor Military Base, VA. This project comprises two primary facilities: a Secure Operations Center and an Administrative & Support Facility, integrated with advanced security features, secure communications systems, and robust support infrastructure. Designed to support 2,500 personnel at peak operational capacity, the completed facility is valued at $180 million. The construction utilizes high-performance materials—including heavy steel, bullet-resistant glazing, and fire-resistive components—combined with cutting-edge technology to ensure a secure and sustainable environment in compliance with federal and military standards.

Leveraging our extensive experience with federal construction and secure facility projects, we have developed a proposal that clearly defines the critical site conditions. Any deviations from these assumptions will be managed through formal change orders. Our proposal is based on the following assumptions:

- Verified site soil bearing capacity meets stringent design requirements for a secure facility.
- No significant underground obstructions were identified during preliminary geotechnical surveys.
- Utility tie-in boundaries and secure access routes are clearly defined in the project drawings.
- The Client is responsible for obtaining all required permits, security clearances, and third-party inspections.

- The Client will provide temporary power and limited security services during construction.
- The facility is designed as a high-security installation incorporating advanced fire suppression and intrusion detection systems per federal guidelines.

Schedule
- Expected Project Award: May 2025
- On-Site Construction Start: July 2025
- Anticipated Project Completion: December 2027

The thirty-month schedule has been carefully planned to allow for comprehensive engineering optimization, timely procurement of high-quality materials, and competitive subcontractor bidding, ensuring that the Fort Valor Command and Control Center is delivered on schedule and meets all security and operational requirements.

Pricing
Lump Sum $180,000,000.00
*Contingency on 'at risk' portion of project is 7.0%
**Fee as a % of total DL is 12%
Additional earnings on positive cash flow position are ~$500,000.00

Competition
There is one other bidder for the Fort Valor Command and Control Center project. XYZ Defense Contractors, recognized for their innovative secure facility designs, have partnered with regional subcontractors for their bid. While XYZ brings considerable experience in federal projects, our proven track record on secure military installations—combined with our streamlined processes and rigorous cost-control strategies—positions us to deliver significant capital savings and superior operational performance. Our bid reflects a robust cost structure and efficiency that provide a competitive advantage in both pricing and quality.

Engineering: (Approximately 5.7% of the project)
The engineering estimate for the Fort Valor Command and Control Center project was developed using our established internal cost structures and

actual labor data for the designated engineering personnel. A detailed, deliverable-based budget has been prepared based on the comprehensive design package provided by the Client. Our fee is structured as a percentage of direct labor costs, ensuring full transparency and alignment with project expectations. Any adjustments due to design modifications will be managed through our formal change order process upon award.

Risks

- The engineering budget was produced rapidly to meet the project's aggressive schedule, potentially exposing us to cost adjustments if unforeseen design complexities arise.
- A significant portion of the engineering work is scheduled during peak workload periods, which could challenge resource availability and necessitate careful project management.

Opportunities:

- Clear design parameters and Client commitment to key aspects of the project ensure that any necessary changes will be Client-funded.
- Leveraging our local expertise and stringent cost control measures provides opportunities to reduce engineering costs and enhance overall project value.
- The conservative project schedule enables proactive resource allocation and effective risk mitigation.
- Our integrated engineering and procurement approach is expected to yield additional efficiencies and cost savings, strengthening our competitive position.

Technology Equipment

The Client has provided a lump sum allocation and detailed specifications for the integrated building systems equipment required for the Fort Valor Command and Control Center. This includes secure HVAC systems, advanced electrical distribution, robust communications networks, and state-of-the-art fire protection systems.

Risks

- Performance of the equipment is dependent on vendor warranties and guarantees; any deficiencies may require costly rework.
- Delays in delivery of critical systems (such as secure HVAC units or fire suppression components) could lead to project cost overruns and schedule penalties. Delivery penalties will be enforced as outlined in our purchase orders.

Opportunities

- Successful installation and commissioning of these systems will set a benchmark for future federal projects, potentially leading to additional business.
- Effective management of the technology equipment phase will improve our cash flow position and further solidify our relationship with the Client.

Utility Equipment: (4.4% of project)

The project includes specifying and procuring a high-efficiency hot oil system, along with essential power distribution and control equipment to support the facility's utility demands.

Risks:

- Due to the accelerated proposal schedule, firm quotes were not secured for these items, which may lead to cost variability upon final vendor selection.

Opportunities:

- All utility equipment will be competitively bid, ensuring market-leading pricing and cost efficiency.

General Contractor Labor & Materials: (37% of project)

Given the scope and specialized requirements of this secure facility, a predominantly Construction Manager (CM) approach will be utilized. A detailed review of construction methodologies will be conducted prior to ground-breaking to optimize labor deployment and material utilization.

Risks

- The competitive subcontract market may see rate fluctuations when construction commences, potentially impacting overall costs.

Opportunities

- All subcontracts will be competitively bid as a lump sum, enabling tight cost control.
- The project benefits from established material pricing and well-defined craft requirements, ensuring a high-quality estimate.
- Effective subcontract retention strategies will bolster our cash flow throughout the project.

Indirects (5.9% of project)
Risks

- Variations in project scope or unforeseen indirect cost drivers may affect overall indirect cost allocations.

Opportunities

- Tight control and regular review of indirect expenses can identify cost-saving opportunities and enhance overall project efficiency.

Overheads: (Group and Corporate Overheads – 1.3% of project)

The estimate includes recovery for group overheads (1% of engineering direct labor plus 0.25% of materials and subcontract costs) along with an additional 1% of materials and subcontract costs for corporate overheads.

Contingency: (4.1% of project)

- A contingency of 4.1% is applied to the total project cost (excluding Client-supplied lump sum allocations).
- This contingency is partially mitigated by the Client's assumption of liability for approved process changes.
- The material take-off is based on high-confidence estimates for commodities such as steel, concrete, pipe, cable, and conduit.
- Rigorous cost controls will be implemented to minimize contingency expenditures and maximize fee efficiency.

Fee: (4.6% of project)

- The fee is structured as 4.6% of the project value, calculated as a percentage of all self-performed labor (Engineering & Field Installation).
- This fee reflects our comprehensive project management, stringent quality control measures, and the value delivered through efficient execution.

Cash Flow

A detailed cash flow plan will be developed for the Fort Valor Command and Control Center project. This plan, based on the federal project progress payment schedule, spend plan, payment terms, and retainage details, is designed to ensure a positive cash position throughout the first two-thirds of the project. Regular updates will be provided to reflect actual expenditures and received funds, facilitating proactive management of working capital.

Terms and Conditions

Since the United States Army has not provided specific terms and conditions for this project, we are submitting our standard federal-compliant terms and conditions as a basis for this proposal. These terms cover indemnification, change orders, payment terms, dispute resolution, and warranties. They will serve as a starting point for negotiation and will be refined during contract discussions to ensure mutual agreement and compliance with federal project requirements.

Project Reviews

To ensure effective oversight, the project will undergo regular reviews led by the Project Manager:

- **Biweekly Project Team Status Meetings:** Monitor daily progress and address emerging issues.
- **Monthly Management Reviews:** Assess overall progress, cost performance, and schedule adherence.
- **Quarterly Executive Summaries:** Provide comprehensive updates to the U.S. Army and corporate leadership.
- **Client Reviews:** Conducted as required by the Army program office.
- Additional reviews may be scheduled at critical milestones or in response to unforeseen challenges.

President Approval

Prior to proposal submission, contract signing, or performing work involving significant risks or exposures, the proposal and contract must be reviewed and approved by the Group President. This ensures that all contractual obligations, risk allocations, and liability limits align with our federal project policies and risk management guidelines.

Limits of Liability

Any contract must include an absolute, legally binding limit on our overall liability to the United States Army. This limit shall not exceed $10,000,000 and must be no less than $5,000,000, unless further approved by the CEO. Such limitations may exclude third-party claims for bodily injury or property damage resulting from our negligence, as permitted by applicable law.

Insurance for the Negligence of Others

No contract shall require us to provide more than $5,000,000 of general liability insurance for another party's negligence, nor shall it automatically grant any party insured status on our insurance programs beyond our own exposure under the contract. Any such requirement must be qualified to reflect our actual risk exposure in accordance with federal guidelines.

Liability for Environmental Services or Professional Opinions

For contracts involving environmental services, significant pollution exposure, hazardous material risks, or those including professional opinions used as a basis for financial decisions, additional contractual safeguards will be implemented. These measures ensure our liability is appropriately limited and risk exposure is clearly defined and managed in compliance with federal standards.

LEVEL III BAR CHART

For a full-size, high-resolution version of this diagram, scan the QR code or visit:
https://projectcontrolsforfederalcontractors.com

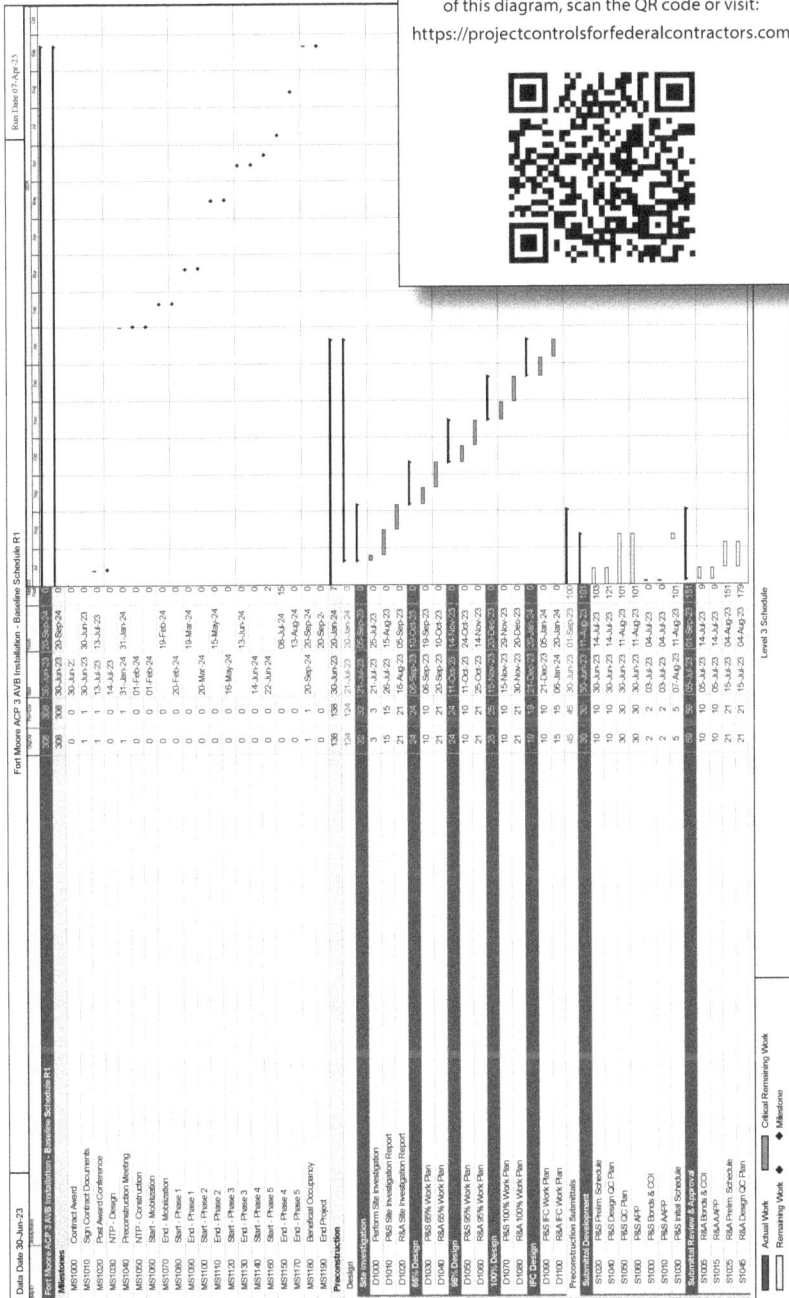

Fort Moore ACP 3 AVB Installation - Baseline Schedule R1

Data Date 30-Jun-23   Run Date 07-Apr-23

Activity ID	Activity Name	OD	RD	Start	Finish
	Fort Moore ACP 3 AVB Installation - Baseline Schedule R1	308	308	30-Jun-23	20-Sep-24
	Milestones	308	308	30-Jun-23	20-Sep-24
MS1000	Contract Award	0	0		30-Jun-23
MS1010	Sign Contract Documents	1	1	30-Jun-23	30-Jun-23
MS1020	Post Award Conference	1	1	13-Jul-23	13-Jul-23
MS1030	NTP - Design	1	1	14-Jul-23	14-Jul-23
MS1040	Preconstruction Meeting	1	1	31-Jan-24	31-Jan-24
MS1050	NTP - Construction	0	0		01-Feb-24
MS1060	Start - Mobilization	0	0		01-Feb-24
MS1070	End - Mobilization	0	0		19-Feb-24
MS1080	End - Phase 1	0	0		20-Feb-24
MS1090	Start - Phase 1	0	0		20-Mar-24
MS1100	Start - Phase 2	0	0		20-Mar-24
MS1110	End - Phase 2	0	0		16-May-24
MS1120	Start - Phase 3	0	0		13-Jun-24
MS1130	End - Phase 3	0	0		14-Jun-24
MS1140	Start - Phase 4	15	15		
MS1150	Start - Phase 5	2	2		13-Aug-24
MS1160	End - Phase 4	0	0		20-Sep-24
MS1170	End - Phase 5	0	0		20-Sep-24
MS1180	Beneficial Occupancy	1	1	20-Sep-24	20-Sep-2
MS1190	End Project	0	0		20-Sep-24
	Preconstruction	138	138	30-Jun-23	30-Jun-24
	Design	124	124	21-Jul-23	31-Jan-24
	Site Investigation	32	32	21-Jul-23	05-Sep-23
D1000	Perform Site Investigation	3	3	21-Jul-23	25-Jul-23
D1010	P&S Site Investigation Report	15	15	26-Jul-23	15-Aug-23
D1020	R&A Site Investigation Report	21	21	16-Aug-23	05-Sep-23
	60% Design	24	24	06-Sep-23	10-Oct-23
D1030	P&S 60% Work Plan	10	10	06-Sep-23	19-Sep-23
D1040	R&A 60% Work Plan	21	21	20-Sep-23	10-Oct-23
	90% Design	24	24	11-Oct-23	14-Nov-23
D1050	P&S 90% Work Plan	10	10	11-Oct-23	24-Oct-23
D1060	R&A 90% Work Plan	21	21	25-Oct-23	14-Nov-23
	100% Design	25	25	15-Nov-23	20-Dec-23
D1070	P&S 100% Work Plan	10	10	15-Nov-23	29-Nov-23
D1080	R&A 100% Work Plan	21	21	30-Nov-23	20-Dec-23
	IFC Design	19	19	21-Dec-23	30-Jan-24
D1090	P&S IFC Work Plan	10	10	21-Dec-23	05-Jan-24
D1100	R&A IFC Work Plan	15	15	08-Jan-24	30-Jan-24
	Preconstruction Submittals	45	45	30-Jun-23	01-Sep-23
	Submittal Development	10	10	30-Jun-23	14-Jul-23
S1020	P&S Prelim. Schedule	10	10	30-Jun-23	14-Jul-23
S1040	P&S Design QC Plan	30	30	30-Jun-23	11-Aug-23
S1060	P&S AFP	30	30	30-Jun-23	11-Aug-23
S1000	P&S AFP	2	2	03-Jul-23	04-Jul-23
S1010	P&S Bonds & COI	2	2	03-Jul-23	04-Jul-23
S1030	P&S Initial Schedule	5	5	07-Aug-23	11-Aug-23
	Submittal Review & Approval	50	50	05-Jul-23	01-Sep-23
S1005	R&A Bonds & COI	10	10	05-Jul-23	14-Jul-23
S1015	R&A AFP	10	10	05-Jul-23	14-Jul-23
S1025	R&A Prelim. Schedule	21	21	15-Jul-23	04-Aug-23
S1045	R&A Design QC Plan	21	21	15-Jul-23	04-Aug-23

Actual Work   ◆ Milestone   Critical Remaining Work   Remaining Work

Level 3 Schedule

Activity ID	Activity Name	OD	RD	Start	Finish	TF
S1035	R&A Initial Schedule	21	21	12-Aug-23	01-Sep-23	151
S1055	R&A QC Plan	21	21	12-Aug-23	01-Sep-23	151
S1065	R&A AFP	21	21	12-Aug-23	01-Sep-23	151
Material Procurement		76	76	06-Sep-23	26-Dec-23	23
LL1000	P&S AVB Systems	5	5	06-Sep-23	12-Sep-23	24
LL1010	R&A AVB Systems	21	21	13-Sep-23	03-Oct-23	35
LL1020	Procure AVB Systems	84	84	04-Oct-23	26-Dec-23	35
Construction		135	135			
MOB1010	Mutual Understanding Meeting	1	1	12-Feb-24	12-Feb-24	0
Mobilization		12	12	01-Feb-24	19-Feb-24	0
MOB1000	Key Personnel Arrive On-Site	8	8	01-Feb-24	13-Feb-24	2
MOB1020	Receive Temp Fence	1	1	13-Feb-24	13-Feb-24	2
MOB1030	Receive Office Trailer	1	1	13-Feb-24	13-Feb-24	2
MOB1040	Receive Rental Equip	1	1	13-Feb-24	13-Feb-24	2
MOB1050	Receive Portable Toilets	1	1	13-Feb-24	13-Feb-24	2
MOB1060	Utility Marking w/ GPR	3	3	13-Feb-24	15-Feb-24	3
MOB1070	Construction Crew Arrives On-Site; Safety Brief	1	1	16-Feb-24	16-Feb-24	0
Phase 1		19	19	20-Feb-24	19-Mar-24	0
CON1000	Install Erosion & Sediment Control Devices - Ph. 1	2	2	20-Feb-24	21-Feb-24	0
CON1010	Set Traffic Controls for Lane Closures - Ph. 1	1	1	22-Feb-24	22-Feb-24	0
CON1020	Layout Outbound Ramp Barrier, Signs & Signals - Ph. 1	1	1	23-Feb-24	23-Feb-24	0
CON1030	Sawcut, Excavate & Dispose Spoils for Barrier Install & EPU Pad - Ph. 1	2	2	26-Feb-24	26-Feb-24	0
CON1040	Prepare Subgrade & Set Outbound Ramp Barrier - Ph. 1	1	1	27-Feb-24	27-Feb-24	0
CON1050	Install Conduits from Outbound Ramp Barrier to EPU Pad Location - Ph. 1	2	2	29-Feb-24	01-Mar-24	0
CON1060	Install Drain Plumbing & Outlet Protection - Ph. 1	1	1	01-Mar-24	01-Mar-24	0
CON1070	Place & Finish Concrete for Wedge Barrier & EPU Pad - Ph. 1	2	2	04-Mar-24	05-Mar-24	0
CON1080	Excavate for Mast Arm, Light Pole, & Wig-Wag Foundations - Ph. 1	2	2	06-Mar-24	07-Mar-24	0
CON1090	Form Curbs - Ph. 1	2	2	08-Mar-24	11-Mar-24	0
CON1100	Place & Finish Concrete Curbs, Mast Arm, Light Pole, & Wig-Wag Foundation - Ph. 1	2	2	12-Mar-24	14-Mar-24	0
CON1110	Trench, Place Conduits & Spacers, Place Concrete Ductbank, & Backfill Pathway (OB Ramp EPU to OB Barriers) - Ph. 1	2	2	15-Mar-24	18-Mar-24	0
CON1120	Assemble & Mount Barrier Signals, Wig-Wags, & Cobra Head Light Fixture - Ph. 1	2	2	15-Mar-24	18-Mar-24	1
CON1130	Set Mast Arm, Wig-Wag Posts, & Light Pole - Ph. 1	1	1	18-Mar-24	18-Mar-24	1
CON1140	Set LED Blank-Out Pole, Install LED Blank-Out Signs - Ph. 1	1	1	18-Mar-24	18-Mar-24	1
CON1150	Sawcut, Install & Seal Safety Loops - Ph. 1	1	1	18-Mar-24	18-Mar-24	0
CON1160	Apply Pavement Markings, Install Stop Here on Red Signs - Ph. 1	1	1	19-Mar-24	19-Mar-24	0
CON1170	Remove Traffic Control for Lane Closures - Ph. 1	1	1	20-Mar-24	20-Mar-24	0
Phase 2		38	38	20-Mar-24	15-May-24	0
CON2000	Set Traffic Control for Lane Closure - Ph. 2	1	1	20-Mar-24	20-Mar-24	0
CON2010	Layout Outbound Barriers, Overwatch Pad, Signs & Signals - Ph. 2	2	2	20-Mar-24	22-Mar-24	0
CON2020	Sawcut, Excavate & Dispose Spoils for Barrier Install Overwatch Pad, & EPU Pad - Ph. 2	2	2	21-Mar-24	22-Mar-24	0
CON2030	Prepare Subgrade & Set Outbound Barriers - Ph. 2	1	1	25-Mar-24	25-Mar-24	0
CON2040	Install Conduits for Outbound Barriers to EPU Pad Location - Ph. 2	1	1	26-Mar-24	26-Mar-24	0
CON2050	Install Sump Pump Discharge Plumbing & Outlet Protection - Ph. 2	1	1	27-Mar-24	27-Mar-24	0
CON2060	Place & Finish Concrete for Wedge Barriers & EPU Pad - Ph. 2	2	2	28-Mar-24	01-Apr-24	0
CON2070	Excavate for Mast Arm, Light Pole, LED Blank-Out Pole & Wig-Wag Foundations - Ph. 2	2	2	29-Mar-24	03-Apr-24	0
CON2080	Form Curbs - Ph. 2	1	1	01-Apr-24	03-Apr-24	0
CON2090	Place & Finish Concrete Curbs, Mast Arm, LED Blank-Out Pole, Light Pole & Wig-Wag Foundations - Ph. 2	2	2	03-Apr-24	08-Apr-24	0
CON2100	Trench, Place Conduits & Spacers, Place Concrete Ductbank & Backfill Pathway (OB Barriers to Overwatch Pad) - Ph. 2	2	2	05-Apr-24	08-Apr-24	0
CON2110	Assemble & Mount Barrier Signals, Wig-Wags, LED Blank-Out Sign & Cobra Head Light Fixture - Ph. 2	2	2	05-Apr-24	08-Apr-24	21
CON2120	Set Mast Arm, Wig-Wag Posts, LED Blank-Out Pole & Light Pole - Ph. 2	2	2	09-Apr-24	11-Apr-24	21
CON2130	Sawcut, Install & Seal Safety Loops - Ph. 2	1	1	09-Apr-24	09-Apr-24	0
CON2210	Sawcut, Install & Seal Wrong Way Loops - Ph. 2	1	1	11-Apr-24	11-Apr-24	21
CON2140	Apply Pavement Markings, Install Stop Here on Red Signs - Ph. 2	1	1	12-Apr-24	12-Apr-24	0
CON2150	Trench, Install Conduit, Place Concrete Ductbank, Backfill & Restore Pathway (OB EPU to WW Loop 2 Location) - Ph. 2	7	7	15-Apr-24	23-Apr-24	0

Actual Work ◆ Remaining Work ████ Critical Remaining Work ◆ Milestone

Level 3 Schedule

Page 2 of 4

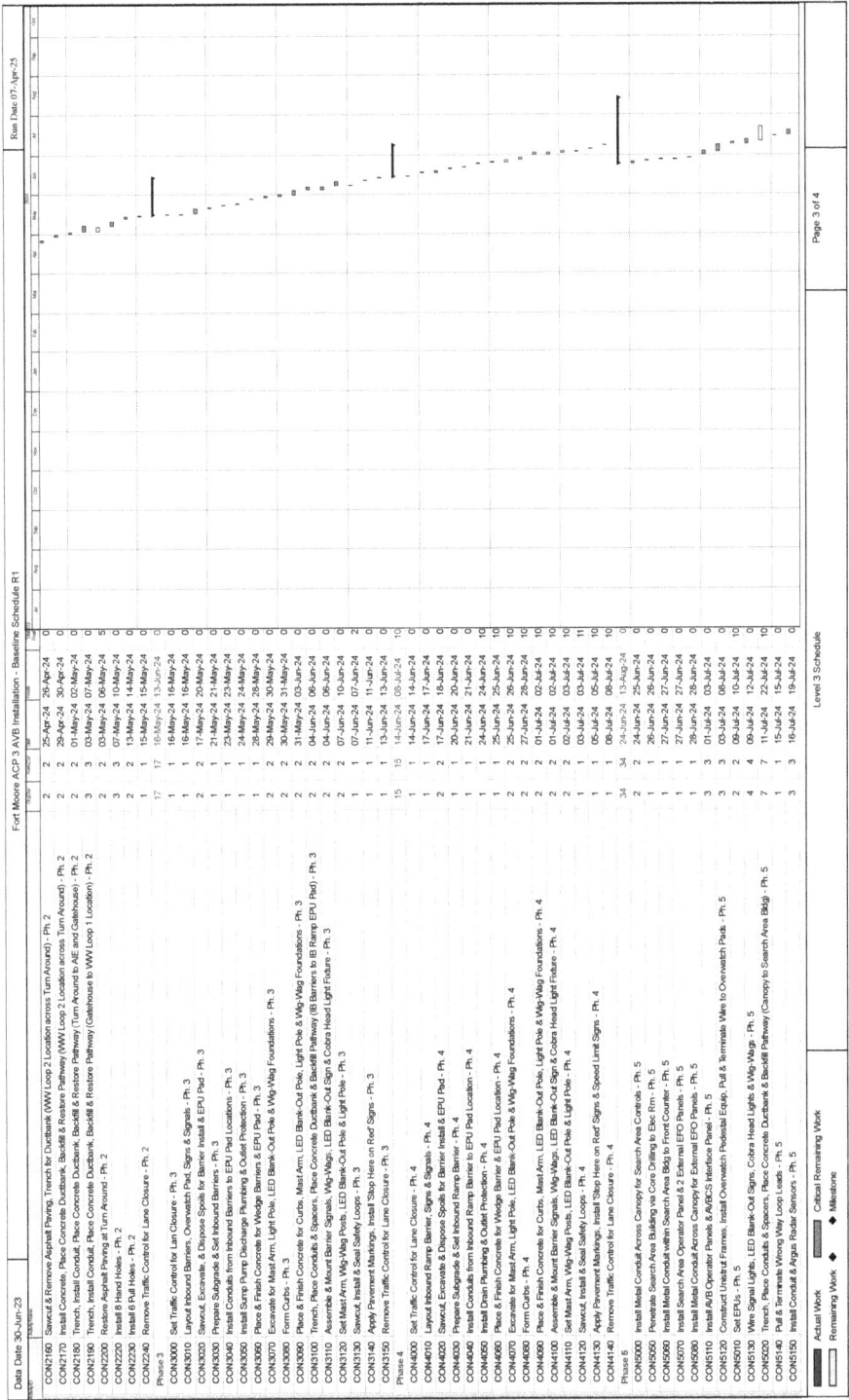

Activity ID	Activity Name	Orig Dur	Rem Dur	Start	Finish	Float
CON2160	Sawcut & Remove Asphalt Paving, Trench for Ductbank (WW Loop 2 Location across Turn Around) - Ph. 2	2	2	25-Apr-24	26-Apr-24	0
CON2170	Install Concrete, Place Concrete Ductbank, Backfill & Restore Pathway (WW Loop 2 Location across Turn Around) - Ph. 2	2	2	29-Apr-24	30-Apr-24	0
CON2180	Trench, Install Conduit, Place Concrete Ductbank, Backfill & Restore Pathway (Turn Around to AIE and Gatehouse) - Ph. 2	2	2	01-May-24	02-May-24	0
CON2190	Trench, Install Conduit, Place Concrete Ductbank, Backfill & Restore Pathway (Gatehouse to WW Loop 1 Location) - Ph. 2	3	3	03-May-24	07-May-24	0
CON2200	Restore Asphalt Paving at Turn Around - Ph. 2	2	2	03-May-24	06-May-24	5
CON2220	Install 8 Hand Holes - Ph. 2	3	3	07-May-24	10-May-24	0
CON2230	Install 6 Full Holes - Ph. 2	3	3	13-May-24	14-May-24	0
CON2240	Remove Traffic Control for Lane Closure - Ph. 2	1	1	15-May-24	15-May-24	0
Phase 3		17	17	16-May-24	13-Jun-24	
CON3000	Set Traffic Control for Lan Closure - Ph. 3	1	1	16-May-24	16-May-24	0
CON3010	Layout Inbound Barriers, Overwatch Pad, Signs & Signals - Ph. 3	1	1	16-May-24	16-May-24	0
CON3020	Sawcut, Excavate, & Dispose Spoils for Barrier Install & EPU Pad - Ph. 3	2	2	17-May-24	20-May-24	0
CON3030	Prepare Subgrade & Set Inbound Barriers - Ph. 3	1	1	21-May-24	21-May-24	0
CON3040	Install Conduits from Inbound Barriers to EPU Pad Locations - Ph. 3	1	1	23-May-24	23-May-24	0
CON3050	Install Sump Pump Discharge Plumbing & Outlet Protection - Ph. 3	1	1	24-May-24	24-May-24	0
CON3060	Place & Finish Concrete for Wedge Barriers & EPU Pad - Ph. 3	2	2	28-May-24	28-May-24	0
CON3070	Excavate for Mast Arm, Light Pole, LED Blank-Out Pole & Wig-Wag Foundations - Ph. 3	2	2	29-May-24	30-May-24	0
CON3080	Form Curbs - Ph. 3	2	2	30-May-24	31-May-24	0
CON3090	Place & Finish Concrete for Curbs, Mast Arm, LED Blank-Out Pole, Light Pole & Wig-Wag Foundations - Ph. 3	2	2	31-May-24	31-May-24	0
CON3100	Trench, Place Conduits & Spacers, Place Concrete Ductbank & Backfill Pathway (IB Barriers to IB Ramp EPU Pad) - Ph. 3	2	2	04-Jun-24	06-Jun-24	0
CON3110	Assemble & Mount Barrier Signals, Wig-Wag, LED Blank-Out Sign & Cobra Head Light Fixture - Ph. 3	2	2	04-Jun-24	06-Jun-24	0
CON3120	Set Mast Arm, Wig-Wag Posts, LED Blank-Out Pole & Light Pole - Ph. 3	2	2	07-Jun-24	10-Jun-24	0
CON3130	Sawcut, Install & Seal Safety Loops - Ph. 3	1	1	07-Jun-24	07-Jun-24	2
CON3140	Apply Pavement Markings, Install Stop Here on Red Signs - Ph. 3	2	2	11-Jun-24	11-Jun-24	2
CON3150	Remove Traffic Control for Lane Closure - Ph. 3	1	1	13-Jun-24	13-Jun-24	0
Phase 4		15	15	14-Jun-24	08-Jul-24	10
CON4000	Set Traffic Control for Lane Closure - Ph. 4	1	1	14-Jun-24	14-Jun-24	10
CON4010	Layout Inbound Ramp Barrier, Signs & Signals - Ph. 4	1	1	17-Jun-24	17-Jun-24	10
CON4020	Sawcut, Excavate & Dispose Spoils for Barrier Install & EPU Pad - Ph. 4	2	2	17-Jun-24	18-Jun-24	10
CON4030	Prepare Subgrade & Set Inbound Ramp Barrier - Ph. 4	1	1	20-Jun-24	20-Jun-24	10
CON4040	Install Conduits from Inbound Ramp Barrier to EPU Pad Location - Ph. 4	1	1	21-Jun-24	21-Jun-24	10
CON4050	Install Drain Plumbing & Outlet Protection - Ph. 4	1	1	24-Jun-24	24-Jun-24	10
CON4060	Place & Finish Concrete for Wedge Barrier & EPU Pad Location - Ph. 4	2	2	25-Jun-24	25-Jun-24	10
CON4070	Excavate for Mast Arm, Light Pole, LED Blank-Out Pole & Wig-Wag Foundations - Ph. 4	2	2	25-Jun-24	26-Jun-24	10
CON4080	Form Curbs - Ph. 4	2	2	27-Jun-24	28-Jun-24	10
CON4090	Place & Finish Concrete for Curbs, Mast Arm, LED Blank-Out Pole, Light Pole & Wig-Wag Foundations - Ph. 4	2	2	01-Jul-24	02-Jul-24	10
CON4100	Assemble & Mount Barrier Signals, Wig-Wag, LED Blank-Out Sign & Cobra Head Light Fixture - Ph. 4	2	2	02-Jul-24	03-Jul-24	10
CON4110	Set Mast Arm, Wig-Wag Posts, LED Blank-Out Pole & Light Pole - Ph. 4	1	1	03-Jul-24	03-Jul-24	10
CON4120	Sawcut, Install & Seal Safety Loops - Ph. 4	1	1	05-Jul-24	05-Jul-24	11
CON4130	Apply Pavement Markings, Install Stop Here on Red Signs & Speed Limit Signs - Ph. 4	1	1	08-Jul-24	08-Jul-24	10
CON4140	Remove Traffic Control for Lane Closure - Ph. 4	1	1	08-Jul-24	08-Jul-24	10
Phase 5		34	34	24-Jun-24	13-Aug-24	
CON5000	Install Metal Conduit Across Canopy for Search Area Controls - Ph. 5	2	2	24-Jun-24	25-Jun-24	0
CON5050	Penetrate Search Area Building via Core Drilling to Elec Rm - Ph. 5	1	1	26-Jun-24	26-Jun-24	0
CON5060	Install Metal Conduit within Search Area Bldg to Front Counter - Ph. 5	1	1	27-Jun-24	27-Jun-24	0
CON5070	Install Search Area Operator Panel & 2 External EPO Panels - Ph. 5	1	1	27-Jun-24	27-Jun-24	0
CON5080	Install Metal Conduit Across Canopy for External EPO Panels - Ph. 5	1	1	28-Jun-24	28-Jun-24	0
CON5110	Install AVB Operator Panels & AVBCS Interface Panel - Ph. 5	3	3	03-Jul-24	03-Jul-24	0
CON5120	Construct Unabilut Frames, Install Overwatch Pedestal Equip, Pull & Terminate Wire to Overwatch Pads - Ph. 5	3	3	03-Jul-24	08-Jul-24	0
CON5130	Set EPUs - Ph. 5	2	2	09-Jul-24	10-Jul-24	0
CON5530	Wire Signal Lights, LED Blank-Out Signs, Cobra Head Lights & Wig-Wags - Ph. 5	4	4	09-Jul-24	12-Jul-24	0
CON5200	Trench, Place Conduits & Spacers, Place Concrete Ductbank & Backfill Pathway (Canopy to Search Area Bldg) - Ph. 5	7	7	11-Jul-24	22-Jul-24	0
CON5140	Pull & Terminate Wrong Way Loop Leads - Ph. 5	1	1	15-Jul-24	15-Jul-24	10
CON5150	Install Conduit & Argus Radar Sensors - Ph. 5	3	3	16-Jul-24	19-Jul-24	0

Legend: Actual Work · Remaining Work · Critical Remaining Work · Milestone

Level 3 Schedule

Page 3 of 4

Data Date 30-Jun-23 — Fort Moore ACP 3 AVB Installation - Baseline Schedule R1 — Run Date 07-Apr-25

Activity ID	Activity Name	Orig Dur	Rem Dur	Start	Finish	Float
CON5160	Install Argus Panel & Wiring to Radar Sensors - Ph. 5	2	2	22-Jul-24	23-Jul-24	0
CON5030	Sawcut Sidewalk, Install Conduit & Patch Concrete - Ph. 5	1	1	23-Jul-24	23-Jul-24	10
CON5040	Site Restoration - Ph. 5	4	4	24-Jul-24	29-Jul-24	10
CON5050	Pull & Terminate Power Wire - Ph. 5	6	6	24-Jul-24	31-Jul-24	0
CON5100	Pull & Terminate Signal Wire - Ph. 5	4	4	01-Aug-24	06-Aug-24	0
CON5170	Start-up & Configure AVBCS - Ph. 5	3	3	08-Aug-24	12-Aug-24	0
CON5180	Contractor Field Test - Ph. 5	1	1	13-Aug-24	13-Aug-24	0
Commissioning		50	56	14-Jun-24	04-Sep-24	11
CX1000	Notify PDC of PVT Date	1	1	14-Jun-24	14-Jun-24	53
CX1010	Perform PVT	2	2	19-Aug-24	20-Aug-24	10
CX1030	Perform O&M Training	1	1	20-Aug-24	20-Aug-24	21
CX1020	Perform Endurance Testing	10	10	21-Aug-24	04-Sep-24	10
Post Construction		88	88	16-May-24	20-Sep-24	0
Closeout Activities		27	27	14-Aug-24	20-Sep-24	0
PC1000	Perform Pre-Final Inspection - Contractor	1	1	14-Aug-24	14-Aug-24	0
PC1020	Perform Pre-Final Inspection - Govt	1	1	14-Aug-24	14-Aug-24	0
PC1010	Correct Pre-Final Punchlist - Contractor	25	25	15-Aug-24	19-Sep-24	0
PC1030	Correct Pre-Final Punchlist - Govt	25	25	15-Aug-24	19-Sep-24	0
PC1040	Pre-Warranty Conference	1	1	05-Sep-24	05-Sep-24	10
PC1050	Perform Final Inspection	1	1	20-Sep-24	20-Sep-24	0
Closeout Submittals		88	88	16-May-24	20-Sep-24	0
Closeout Submittal Development		78	78	16-May-24	05-Sep-24	10
S1070	P&S PVT Plan	10	10	16-May-24	30-May-24	54
S1080	P&S O&M Documentation	10	10	16-May-24	30-May-24	64
S1050	P&S Warranty Mgmt Plan	15	15	09-Jul-24	29-Jul-24	24
S1110	P&S Completion Photos	5	5	14-Aug-24	20-Aug-24	22
S1120	P&S As-Built Drawings	10	10	14-Aug-24	27-Aug-24	3
S1130	P&S Interim DD1354	1	1	14-Aug-24	14-Aug-24	0
S1100	P&S PVT Report	10	10	21-Aug-24	04-Sep-24	12
S1140	P&S Final DD1354	2	2	05-Sep-24	05-Sep-24	0
Closeout Submittal Review & Approval		113	113	31-May-24	20-Sep-24	0
S1075	R&A PVT Plan	21	21	31-May-24	20-Jun-24	80
S1085	R&A O&M Documentation	21	21	31-May-24	20-Jun-24	92
S1055	R&A Warranty Mgmt Plan	21	21	30-Jul-24	19-Aug-24	32
S1135	R&A Interim DD1354	21	21	15-Aug-24	04-Sep-24	0
S1125	R&A As-Built Drawings	21	21	28-Aug-24	17-Sep-24	3
S1145	R&A Final DD1354	14	14	07-Sep-24	20-Sep-24	0

Actual Work   Critical Remaining Work   Remaining Work   ◆ Milestone

Level 3 Schedule   Page 4 of 4

LEVEL II BAR CHART

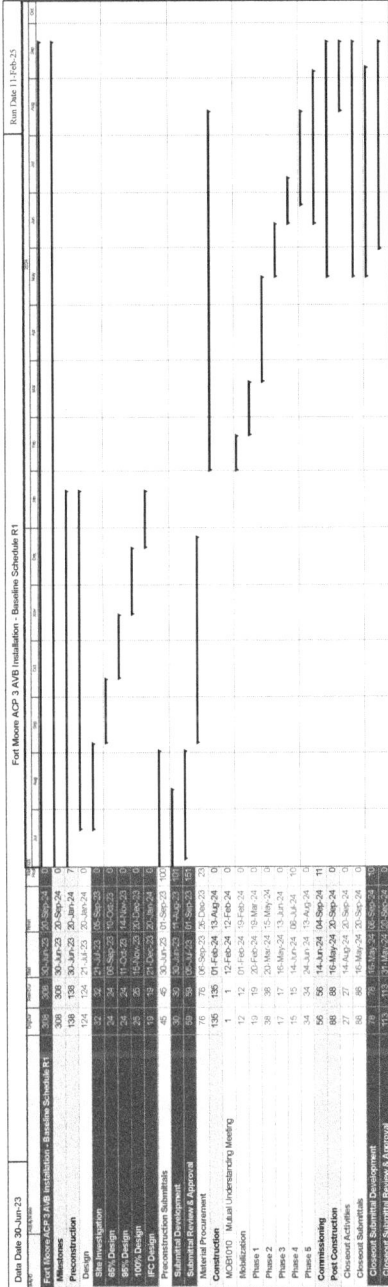

For a full-size, high-resolution version of this diagram, scan the QR code or visit: https://projectcontrolsforfederalcontractors.com

Data Date 30-Jun-23 — Run Date 11-Feb-23

Fort Moore ACP 3 AVB Installation - Baseline Schedule R1

Activity	Orig Dur	Rem Dur	Start	Finish	TF
Fort Moore ACP 3 AVB Installation - Baseline Schedule R1					
Milestones	303	303	10-Jun-23	20-Sep-24	0
Preconstruction	303	303	30-Jun-23	20-Jan-24	0
Design	138	138	30-Jun-23	20-Jan-24	7
Design	124	124	21-Jul-23	30-Jan-24	0
Site Investigation	32	32	21-Jul-23	05-Sep-23	
65% Design	24	34	06-Sep-23	10-Oct-23	
95% Design	24	24	11-Oct-23	14-Nov-23	
100% Design	25	25	15-Nov-23	20-Dec-23	
IFC Design	19	19	21-Dec-23	20-Jan-24	
Preconstruction Submittals	45	45	30-Jun-23	01-Sep-23	100
Submittal Development	30	30	30-Jun-23	11-Aug-23	155
Submittal Review & Approval	59	59	06-Jul-23	01-Sep-23	155
Material Procurement	78	76	06-Sep-23	25-Dec-23	23
Construction	135	135	01-Feb-24	13-Aug-24	23
MOBIDYD Mutual Understanding Meeting	1	1	12-Feb-24	12-Feb-24	
Mobilization	12	12	01-Feb-24	19-Feb-24	
Phase 1	19	19	20-Feb-24	19-Mar-24	
Phase 2	38	38	20-Mar-24	15-May-24	
Phase 3	17	17	16-May-24	13-Jun-24	
Phase 4	15	15	14-Jun-24	05-Jul-24	10
Phase 5	34	34	24-Jun-24	13-Aug-24	10
Commissioning	56	56	14-Jun-24	04-Sep-24	11
Post Construction	88	88	16-May-24	20-Sep-24	0
Closeout Activities	27	27	14-Aug-24	20-Sep-24	
Closeout Submittals	88	88	16-May-24	20-Sep-24	
Closeout Submittal Development	78	78	16-May-24	06-Sep-24	10
Closeout Submittal Review & Approval	113	113	26-May-24	20-Sep-24	0

Summary — Level 2 Schedule

JOBSITE INSPECTION CHECKLIST

Safety Items

_____ Clean drinking water with plenty of cups

_____ 29CFR/EM385 present on jobsite

_____ Power tagged with safety color of the month

_____ Power sources tagged reflecting re-certification within the last month

_____ Fire extinguishers staged and checked within the last month

_____ Welders or light plants not used as power source

_____ GFCI in use

_____ All rebar is capped

_____ Scaffolding has handrails, toe boards, and bracing

_____ Project site roped off

_____ All vehicles have wheels chocked

Housekeeping

_____ Jobsite should be clean for inspection

_____ Conex boxes and kits neat and orderly

_____ Vehicles clean and pre-started

_____ Excess material removed from the jobsite

_____ Professional project sign with current info

Crewmembers

_____ Must know who is the project safety representative

_____ Must know the safety color of the month

_____ Must wear hard hats at all times

_____ Must know the QC inspector on jobsite

_____ Must know the information on the CASS for the current and next activities

_____ Must know timekeeping procedures

_____ Strict adherence to daily routine

_____ Assistant supervisors designated

Project Management

_____ Bar charts and CASS reflect actual status

_____ Project package is complete and in use by Project Manager

_____ As-Builts (Redlines) drawings are up to date

_____ BM reflects accurate material status

_____ Project logs up to date

_____ Direct labor reports up to date

APPENDIX 4-2

ACTIVITY HAZARD ANALYSIS (AHA)

US Army Corps of Engineers (USACE)

For use of this form, see EM 385-1-1; the proponent agency is CESO

Purpose: The Activity Hazard Analysis (AHA) is a tool used in the Risk Management Process. Risk management is a business process that includes the identification, assessment, and prioritizing of risks, followed by coordinated and economical application of resources to minimize, monitor, and control the probability and/or impact of unfortunate events to an acceptable level.

The overall residual risk assessment code (RAC) must be communicated and accepted by the proper approval authority before beginning the activity.

AHAs must be provided to, and reviewed by, all involved employees prior to starting the task. Each employee must document their review with a signature in the last section of form.

Activity:	Mobilization & Site Set-up	**Date:**	06/15/2025
Location:	100 Valor Way, Fort Valor Military Base, VA 22301	**Overall RAC** *(Use highest code)*	M
Prepared by:	Maria Rodriguez (SSHO)		
Reviewed by:	See Employee Document Sheet		
Notes:	All personnel must review this AHA and adhere to established safety procedures and PPE requirements prior to commencing work.		

Risk Assessment Code (RAC) Matrix

E = Extremely High Risk H = High Risk M = Medium Risk L = Low Risk		Probability				
		Frequent	Likely	Occasional	Seldom	Unlikely
SEVERITY	Catastrophic	E	E	H	H	M
	Critical	E	H	H	M	L
	Moderate	H	M	M	L	L
	Negligible	M	L	L	L	L

Job Steps	Hazards (Recognized/Anticipate)	Controls (Actions to Eliminate or Minimize Hazards)	Residual RAC
Unloading Materials & Equipment	- Slips, trips, and falls due to wet or uneven surfaces. - Collision with moving equipment. - Manual handling injuries.	- Ensure staging areas are clear and free of debris. - Use non-slip mats and proper signage. - Utilize certified forklift operators and designate travel routes. - Implement proper lifting techniques.	M
Setting Up Temporary Facilities	- Falls from heights during setup of site office, storage, and fencing. - Manual handling risks. - Exposure to weather elements.	- Employ fall protection systems and ensure proper use of ladders and scaffolds. - Enforce safe manual handling procedures through training. - Conduct periodic inspections of temporary structures.	L

Equipment	Training	Inspection
Forklifts, ladders, scaffolds, temporary fencing, non-slip mats, and additional security screening equipment.	Certified forklift operation, ladder safety, manual handling procedures, and security protocol training specific to military base requirements.	Daily site safety inspections conducted by the Superintendent.

Involved Personnel:
- John Davis – Senior Project Coordinator (Installer)
- David Nguyen – Superintendent
- Maria Rodriguez – SSHO Manager
- Jessica Miller – Project Manager

PPE required:
Hard hats, high-visibility vests, non-slip footwear, gloves, safety glasses, and, where applicable, fall protection harnesses along with any additional security gear mandated by Fort Valor Military Base protocols.

Approval Authority (digital signature)	
Printed Name: Maria Rodriguez	**Printed Title**: SSHO Manager
Digital Signature: Maria Rodriguez	**Date**: 06/15/2025

EMPLOYEE DOCUMENTATION SHEET

AHAs must be provided to and reviewed by all employees involved prior to starting the task.

Each employee must document their review with a signature.

Name	Occupation/ Designation (Job Title)/Competent or Qualified Person	Signature	Date
John Davis	Senior Project Coordinator	John Davis	06/15/2025
David Nguyen	Superintendent	David Nguyen	06/15/2025
Maria Rodriguez	SSHO Manager	Maria Rodriguez	06/15/2025
Jessica Miller	Project Manager	Jessica Miller	06/15/2025

ENVIRONMENT PROTECTION PLAN (EPP)

[Project/Reference Photo]

Produced By
[Company Name
Company Address
Company City, State, Zip]

Produced For
[Owner's Name
Project Name
Contract Number]

Location
[Project Address
Project County, State]

Table of Contents

The information provided Appendix 4-3 regarding the Environmental Protection Plan (EPP) reflects practices commonly found in industry-standard construction specifications. For project-specific requirements, consult your contract documents or contact the local contracting office for guidance. Additionally, ensure your project complies with all applicable state and local environmental regulations.

STORMWATER POLLUTION PREVENTION PLAN (SWPPP)

[Project/Reference Photo]

Produced By

[Company Name

Company Address

Company City, State, Zip]

Produced For

[Owner's Name

Project Name

Contract Number]

Location

[Project Address

Project County, State]

Contents

Appendix 4-4 outlines common components of Stormwater Pollution Prevention Plans (SWPPPs) based on Environmental Protection Agency (EPA) guidance. To ensure full compliance, review the specific state and local stormwater regulations applicable to your project.

QUALITY CONTROL PLAN (QCP)

[Project/Reference Photo]

Produced By
[Company Name
Company Address
Company City, State, Zip]

Produced For
[Owner's Name
Project Name
Contract Number]

Location
[Project Address
Project County, State]

Quality Control Plan (QCP)

A construction QCP is submitted in a three-ring binder that includes a table of contents, with major sections identified with tabs, with pages numbered sequentially, and that documents the proposed methods and responsibilities for accomplishing commissioning activities during the construction of the project.

Contents

The information provided in Appendix 4-5 regarding the Quality Control Plan (QCP) reflects practices commonly found in industry-standard construction specifications. For project-specific requirements, consult your contract documents or contact the local contracting office for guidance.

JUNE STATUS UPDATE

[Company Logo]

PROJECT INFORMATION	
Project Information	
General Contractor	Horizon Builders, Inc.
Project	Fort Valor Command and Control Center
Location	Fort Valor Military Base, VA
Contract	FVC-2025-001

Schedule Data	
Schedule	Baseline
Report Date	06/30/2025
Data Date	06/30/2025
Start Date	07/01/2025
CCD Date	12/31/2027
Finish Date	12/31/2027
On-Schedule	Yes

Software Used

Primavera P6 has been used in the creation of this schedule. This and all future schedules will be done with Primavera P6 Software.

Executive Summary

During June 2025, field operations at Fort Valor Military Base progressed with minimal disruptions. Notably, the exterior secure cladding installation on the Secure Operations Center advanced according to plan despite minor weather-related delays. Critical path activities, such as foundation work and grade corrections, were completed on schedule, ensuring that subsequent tasks—including utility tie-ins and secure access system installations—remain unaffected.

KEY UPDATES

Project Milestones

The following milestones are included in the project schedule.

Activity ID	Activity Name	Start/Finish Date	Total Float
MILE-1000	NTP	07/01/2025	0
MILE-1010	Substantial Completion	10/31/2027	6
MILE-1020	Contract Completion	12/31/2027	0

In-Progress Activities

The following activities are currently in progress.

Activity ID	Activity Name	Start	% Complete
A-102	Site Mobilization	07/01/2025	50%

Activities Completed This Month

The following activities have been completed this schedule update.

Activity ID	Activity Name	Completion Date
A-100	Temporary Site Setup	06/10/2025

Activities Not Started This Update

The following activities were previously anticipated to start this month, but work did not progress according to the schedule.

Activity ID	Activity Name	Reason for Delay
A-103	Utility Tie-in	Awaiting final design confirmation

Work Scheduled to Start in the Next Update

The following activities are anticipated to start next month.

Activity ID	Activity Name	Start	Finish
A-104	Structural Steel Erection	07/05/2025	07/20/2025

Current Critical Path

The following activities comprise the longest path of the schedule.

Activity ID	Activity Name	Start	Finish	Total Float
A-105	Foundation & Grade Work	06/01/2025	6/20/2025	0 days

The current critical path is defined by the foundation and grade work, completed on schedule. This ensures that subsequent tasks—including secure cladding installation and utility tie-ins—remain unaffected.

Current Delays

The following delays are currently on-going:

No significant delays reported; minor weather-related interruptions were effectively mitigated.

Anticipated Future Delays

The following delays are anticipated to impact the project in the upcoming months:

A potential delay in finalizing utility tie-in designs may impact the start of related work. Mitigation measures include expedited review sessions with the design team.

Pending Items

The following items are currently pending:

Response to RFI regarding bracket spacing adjustments for the façade panels is still pending.

SCHEDULE CHANGES

New/Deleted Activities

The following activities have been added or deleted in this update.

Activity ID	Activity Name	Added/ Deleted	Reason
A-106	Additional Cladding Repair	Added	Field adjustment due to site conditions

Logic Changes

The following logic changes have been made this update to better reflect current project sequencing.

Activity ID	Activity Name	Pred	Action	Reason
A-107	Resequencing of material deliveries	A-100	Modified	Optimize staging workflow

Duration Changes

The durations for the activities listed below have been updated this month to reflect more accurate timeframes for performing work.

Activity ID	Activity Name	Orig. Dur.	New Dur.	Reason
A-108	Temporary facility setup	5 days	4 days	Improved work efficiency observed

Lag Changes

The following lag changes have been made to activity logic in this update.

Activity ID	Activity Name	Pred. Lag.	Succ. Lag	Reason
A-109	Utility connection	2 days	0 days	Adjusted minor contractor delay

Cost Changes

The dollar values for the activities below have been re-allocated this month.

Activity ID	Activity Name	Orig. Budgeted Costs	New Budgeted Costs	Reason
A-110	Site preparation	$1,000,000.00	$1,100,000.00	Revised earthwork estimates

Actual Start/Finish Changes

Revisions to the actual start of the below activities have been made this schedule update.

Activity ID	Activity Name	Prev. Actual Start	New Actual Start	Reason
A-111	Material staging	06/05/2025	06/06/2025	Delayed material delivery

Out of Sequence Logic Changes

The Activities listed below have been performed out of sequence compared to how the project was originally scheduled.

Activity ID	Activity Name	Pred	Action	Reason
A-112	Landscaping	A-105	Resequenced	Adjusted to accommodate phased grading

SITE LAYOUT OBJECTIVES CHECKLIST

A. Overall Goals
1. Economy of operation and minimum resource waste
- Fewer Men
- Less Equipment
- Less Material Waste
- Less Time
- Less Management
- Less Money

B. Importance of Effective Site Organization
1. Raises both individual and overall crew productivity. Optimizes efficiency periods while minimizing delays, movement, and idle time resulting from process interruptions.
2. Saves total project man-days/durations
3. Facilitates faster project delivery times
4. Minimizes machinery utilization and return visits
5. Decreases waste, theft, breakage and spoilage of stored supplies
6. Less wear and tear on finished and installed work surfaces
7. Safer work
8. Improves site security
9. Reduces supervisor's daily time spent on minute production decisions, freeing up time for more significant tasks
10. Better impression on Customer and/or end user

C. Specific Objectives
1. Unimpeded workflows
2. Shortest, simplest transport paths
3. Safeguards project resources

4. Ease of receiving, inventorying and staging materials, parts and equipment
5. Optimizes the Project Manager's ability to monitor and control work site operations:
   - Visibility of work and staging areas
   - Ease of access to work and materials
   - Ease of inspection of work and materials
6. Maintain operational flexibility as the character of work progresses from rough and heavy trades to finishing trades
7. Coping with site constraints
   - Dust, temperature, wind, sun, rain, humidity
   - Terrain, surface and subsurface condition
   - Scope and class of work
   - Other interferences and operations

D. Site Organization Checklist
   1. Job Work and traffic flows
      - Exterior access to site
      - Interior traffic and cross traffic
      - Material bottlenecks
      - Work access bottlenecks
      - Trades and processes working through other trades.
      - Analysis of work sequences
      - Plot movements of both temporary and installed materials from staging area and delivery yards to installation points
      - Consider methods used by each trade for:
        - Delivery of materials
        - Erection of materials
        - Use of scaffolding, hoists, elevators
        - Work areas required for proper usage of mobile equipment
      - Consider benefit or hindrance caused by growing mass of installed work
      - Continuous flows have priority (shorter) paths over occasional flows

- Heavier, bulkier, more delicate components have closer location priority than lighter, less bulky, less delicate items
- Equipment must have access to all working points and be provided adequate work-bay maneuvering room

2. Yard spaces and areas for prefabrication of components
   - Temporary structures and assemblies:
     - concrete forms
     - scaffolding
     - hoists
     - lagging, shoring, braces, etc.
     - saw tables, jigs
     - pine fabrication tables, jigs
   - Installed assemblies:
     - rebar
     - embedded items
     - piping and plumbing
     - lumber framing
     - steel framing
     - sheet and misc. metals
     - electrical duct, panels and wire harnesses
     - masonry
     - tile
     - cabinetry and finish carpentry
     - pile driving
     - dewatering
     - instrumentation
     - precast framing, lintels, walls and slabs
3. Materials staging and storage
   - should be coordinated simultaneously with task patterns, operational assessments, and positioning of facility equipment and areas
   - delivery points for inspection receiving and counting
   - store by trade in separate areas

- stockpiling arrangement
  - first used, last used material locations
  - palletized, stacked, sheltered and raised off ground
  - drainage
  - unpacking and assembly space
  - aisle and loader access
  - storage space (i.e. pipe, steel, rebar, joists, electric duct, poles)
  - space for phased delivery of perishable or long lead-time materials
- protection against:
  - environmental deterioration
  - handling and transporting
  - theft, losses and shrinkages
- box trailers for small and valuable parts
- ground treatment, platforms, corduroying mats, etc., for heavy loads on soft ground
- aggregate stockpiles

4. Equipment and field maintenance
   - parking
   - maintenance pads, shops, and shelters
   - FOG/POL storage
   - fueling stations
   - parts van, box trailers
   - tires
   - attachments
   - low boys
   - loading/unloading ramps
   - dust control
   - batch and mix plants
     - compressed air
     - water
     - power
     - foundations
     - bias, stockpiles, surge piles

- loading, scalping, screening, crushing, conveying, washing, stockpiling, reloading, hauling
- cooling or heating aggregate
- method of delivery
- method of placing
- proximity to aggregate sources
- proximity to job site

5. Haul road and pits
   - haul economy
     - elevation (minimum climb, level)
     - length (minimum haul time)
     - grades (minimum shift downs)
     - surfaces (rolling resistance)
     - temporary drainage
     - haul road maintenance

6. Temporary utilities
   - electric power distribution points, lighting
   - water points and water storage
   - potable water
   - sanitary toilets
   - temporary storm drainage
   - fire lines
   - fuel
   - office space
   - employee bulletin boards
   - survey benchmarks and reference points
   - watchman's shack
   - borrow and material pits

7. Security and safety
   - gates, fences, and locked storage
   - barricades and safety lights
   - watchman
     - shack
     - telephone
     - security lighting

8. People
   ▪ personnel report in
   ▪ safety and administrative instruction
   ▪ parking
   ▪ visitors
   ▪ customer offices
   ▪ first aid

APPENDIX 6
SAMPLE DISTRIBUTION LOG

Project Name: Fort Valor Command and Control Center
Project Number: FVC-2025-001
Client/Owner: US Army, Fort Valor
General Contractor: Horizon Builders, Inc.
Location: Fort Valor Military Base, VA 22301
Date: 06/25/2025

Reminders

- Ensure all critical documents are logged immediately upon distribution.
- Keep digital and physical copies (if applicable) for recordkeeping.
- Use unique identifiers for RFIs, submittals, and other documents to avoid confusion.
- Regularly update the log to maintain accurate records.

Date Sent	Document Name/Number	Revision	Sent By	Recipient(s)	Method (email, hard copy, etc.)	Remarks
06/25/2025	Submittal Log	Rev 1	John Doe	Project Engineer, QC Manager	Email	Approved by QC Manager
06/25/2025	RFI #001	Rev 0	Jane Smith	Architect, Structural Engineer	Procore	Awaiting Response
06/25/2025	Updated Drawings (Set A)	Rev 2	Tom Brown	Site Superintendent, Foreman	Hard Copy	Issued for Construction
06/25/2025	Safety Plan	Rev 1	QC Manager	USACE Project Manager	Email	Approved
06/25/2025	Pay Application #3	Rev 0	Accounts Payable	Owner, Contract Administrator	Email	Approved

APPENDIX 6-1

BOM: REWORK

Material information

Part #:	7000-1245632	Part Description:	Door Frame Assembly
Original Lot #:	ZZ45221	Rev:	01
Quantity Involved:	5	Date:	04/25
New Lot #:	ZZ45221-1A		
Source of Rework:	☐ NCMR (NCMR #:_____) ☐ DCO (DCO #: _____) ☐ Other (Describe)		

Proposed Rework Instruction

Delivered door frame assemblies are of knock down frame type and style. Specifications and approved submittals indicate a welded frames assembly style.

Verification/Re-inspection Instructions

Verify and inspect in compliance with project documents, and acceptable industry standards.

Risk Assessment

Severity	Probability	Risk Level
Low	Low	Low

Are there any potential adverse effects of the proposed work?	
If yes, explain:	

Approvals (date, sign)

Quality
Approval
Print Name	Signature	Date

Engineering
Approval
Print Name	Signature	Date

Customer
Approval
(if applicable)
Print Name	Signature	Date

Other
Approval
(if applicable)
Print Name	Signature	Date

Rework (attach re-inspection results when completed, as applicable) ☐ N/A

Performed By
Print Name	Signature	Date

Inspected by
☐ Attach
inspection
results
Print Name	Signature	Date

Inspection Results

Quantity Accepted:		Quantity Rejected:	

Approval signatures:

Performed By
Print Name	Signature	Date

Reviewed &
Approved by
Print Name	Signature	Date

Rev History

Rev	Effective Date	By	Change Description

WARRANTY REQUEST FORM

Requested by:	Jessica Miller		Date:	01/15/2028
Requester is responsible to file a completion notice to the QA department when work is completed.				

Customer Name:	United States Army, Fort Valor	Job No.:	RCP-2025-001
Address:	250 Command Center Blvd, Suite 200		
City, State:	Fort Valor, VA	Zip:	22301
Customer Contact:	Jane Smith	Phone Number:	(804) 555-0202

Describe the cause of the warranty and work to be performed:
During a routine post-occupancy inspection of the Secure Operations Center, it was observed that several automatic sprinkler heads in the north corridor failed to activate during testing. Preliminary investigation indicates that the issue may be due to improper calibration of the sprinkler control valves or defective sprinkler heads. The warranty request is for a full evaluation of the sprinkler system, recalibration of control valves, replacement of any defective sprinkler heads, and subsequent retesting to ensure compliance with NFPA standards and applicable military fire safety requirements.

All above information must be provided including a detailed explanation before approval will be granted.

Approvals		Sales Use Only	
QA Manager:	Robert Thompson	**Warranty No.:**	WRNT-2028-001
Credit:	Approved	**Customer No.:**	1001
Sales:	Jessica Miller	**Market Code:**	Federal
*Goodwill:	None		
**Over $2,500			

All goodwill warranties must have approval by an officer of the Company or a functional manager. The president of the Company must approval all goodwill warranty work over $1,000.00.

**The president or an officer of the Company must approve all warranty work over $2,500.00.

Explanation for rejection:
[If applicable, provide explanation here; otherwise, leave blank.]

Attach to all copies of production order (work order).

BOM: ADD-ONS

Assembly Name: *Steel Door Frame Assembly*

Assembly Number: *7000-1245632*

Assembly Revision *01*

Date of Approval: *04/21/2025*

Total Piece: *5*

Total Cost: *$6,556.00*

BOM Level	Raw Materials	Assemblies & Subassemblies	Part Number	Part Name	Part Description	Unit Cost	Quantity	Total Cost	Procurement Details
1	Steel	Metals	123	Frame	ACE Manufacturing	$1,200.00	5	$6000.00	Freight shipping
2	Primer	Paints/Coatings	654	Frame	ACE Manufacturing	$500.00	1	$500.00	Incl. w/ frames
3	Non-shrink Grout	Misc.	789	Features	ACE Manufacturing	$6.00	1	$6.00	Local pick up
						Total	5	$6,556.00	

APPENDIX 7
NOTIFICATION FOR READY FOR CONSTRUCTION DOCUMENTS

[Document on Company Letterhead]

06/25/2025

John Doe
ABC Electric, Inc.
1234 Electric Ave, Suite 300
Richmond, VA 22301
Email: johndoe@abcelectric.com

Subject: Notification of Issued for Construction (IFC) Drawings for Riverfront Corporate Campus Development

Dear John Doe,

We are pleased to inform you that the "Issued for Construction" (IFC) drawings for the Fort Valor Command and Control Center project have been finalized and are now available for your scope of work. These documents represent the approved design and specifications to be followed during construction on Fort Valor Military Base.

Key Details
- **Project Name:** Fort Valor Command and Control Center
- **Project Location:** 100 Valor Way, Fort Valor Military Base, VA 22301
- **Drawing Reference Numbers:** A-15, S-08, E-12
- **Scope of Work:** Electrical installation and related secure systems for the Secure Operations Center

Action Required

- **Review IFC Drawings**: Please access and thoroughly review the IFC drawings available at our shared drive (http://documents.horizonbuild-ers.com/IFC) or contact our office to obtain physical copies.

- **Confirmation**: Kindly acknowledge receipt and confirm your understanding of the IFC drawings by replying to this notification by 06/28/2025.

- **Implementation**: Ensure that all work performed aligns strictly with the IFC drawings. Any discrepancies or questions should be directed to our Project Manager, Jessica Miller, at (312) 555-0303 or jmill@horizonbuilders.com.

Issued for Construction Stamp

To signify that these drawings are approved for construction, they have been stamped as follows:

ISSUED FOR CONSTRUCTION
DATE: INITIAL:

The "Issued for Construction" stamp indicates that the drawings have been reviewed and approved by the relevant authorities and are now the official documents to be used during the construction phase. Please ensure that all team members involved in your work are informed and have access to the IFC drawings. Adherence to these documents is critical for the successful and timely completion of the project.

Importance of IFC Drawings

IFC drawings are the final set of plans that include all necessary details, specifications, and instructions required for construction. They serve as the legal documents guiding the construction process and ensure that all parties are working from the same approved designs.

If you have any questions or require further clarification, please do not hesitate to contact Jessica Miller at (312) 555-0303 or jmill@horizonbuilders.com.

Thank you for your attention to this matter.

Sincerely,

James Anderson

James Anderson
Project Manager
Horizon Builders, Inc.
(312) 555-0101
janderson@horizonbuilders.com

APPENDIX 8-1
REQUEST FOR INFORMATION (RFI)

RFI Number:	RFI-001	Page:	1 of 1	Route to	Initial	Date
Limit Number:	LMT-01	Project Number:	FVC-2025-001	Installer	JD	06/21/2025
Submitted By:	Jessica Miller	Date:	06/21/2025	PM	JM	06/21/2025
				Eng.	DL	06/21/2025
				GC	DN	06/21/2025
				OBO	JS	06/21/2025

Description of and reason for request (include drawing and sheet numbers and attached drawings as necessary for description):
During the installation of the exterior secure cladding panels on the north elevation of the Secure Operations Center, discrepancies were noted between the field conditions and the approved construction documents. Specifically, on Drawing A-15 (Sheet 3), the mounting bracket spacing is specified as 48 inches. However, field measurements indicate a spacing of 52 inches, likely due to minor variances inherent to secure construction on Fort Valor Military Base. This RFI is submitted to request clarification on whether the design intent mandates strict adherence to the 48-inch spacing or if a field adjustment is acceptable given the observed variance. Relevant drawings and field measurement documentation are attached for review.

Jessica Miller	*Jessica Miller*	06/21/2025
Name of Requestor	**Signature**	**Date**

Clarification from Owner's Representative/General Contractor:		
The Owner's Representative clarifies that while the design intent specifies a 48-inch bracket spacing as the minimum requirement, a variance of up to 4 inches is acceptable provided that the overall secure cladding alignment and structural load calculations remain within acceptable limits. If necessary, a revised detail drawing will be issued.		
Jane Smith	*Jane Smith*	06/22/2025
Name of Approving Official	**Signature**	**Date**

REQUEST FOR INFORMATION (RFI) SUBMITTAL LOG

RFI Number	Description	Spec Section	Drawing Number	Date to GC	Date Returned	Approved/ Denied
RFI-001	Exterior Secure Cladding Panel Spacing Variance	07 24 00	A-15	06/21/2025	06/22/2025	Approved
RFI-002	Clarification on Secure HVAC Duct Routing	23 05 00	M-03	06/22/2025	06/23/2025	Approved
RFI-003	Utility Connection Detail Discrepancy	22 10 00	E-12	06/23/2025	06/24/2025	Denied

APPENDIX 9

CONTRACTOR'S QUALITY CONTROL REPORT (QCR) DAILY LOG OF CONSTRUCTION

Report Number	QCR-2025-001
Date	06/20/2025
Project	Fort Valor Command and Control Center
Contract Number	RCP-2025-001

General Contractor	Weather
Horizon Builders, Inc.	No delay; Temp Min 55°F / Max 70°F; Precipitation 0.00 in; Wind 7 MPH

QC Narratives

Today's inspection confirmed that all temporary facilities and security measures are in place per project specifications and federal security standards. The secure cladding installation on the Secure Operations Center was observed for compliance. A minor deficiency in the placement of temporary security signage at the main entrance was identified and corrected within the day. All work is proceeding as scheduled with no critical deviations noted.

Prep/Initial Dates

Mobilization & Site Security Setup initiated: 06/15/2025

Activity Start/Finish

Mobilization & Site Security Setup: Start 06/15/2025, Finish (anticipated) 06/20/2025

QC Requirements

- Daily site inspections
- Verification of temporary facility and secure utility coordination
- Compliance with safety, security protocols, and project specifications

Issued QA/QC Deficiencies

Temporary security signage not correctly located at the main entrance

Corrected QA/QC Deficiency

Signage repositioned to the specified secure location on 06/17/2025

General Contractors on Site

Horizon Builders, Inc. personnel (installation crew, supervisors, QC staff)

Labor Hours

Employer	Labor Classification	Number of Employees	Hours Worked
Horizon Builders, Inc.	General Labor	5	40
Horizon Builders, Inc.	Skilled Trades	3	24
Horizon Builders, Inc.	Site Supervisor	1	8
Total:		**9**	**72**

Total HRS Worked to Date:	72	HRS Worked MTD:	72

Equipment Hours

Serial Number	Description	Idle Hours	Operating Hours
EQ-001	Mobile Crane	2	6
EQ-002	Forklift	0	8
Total:		**2**	**14**

Total Operating HRS to Date:	14	Idle HRS MTD:	2	OPR HRS MTD:	14

Mishap Reporting

No mishaps reported.

General Contractor Certification: On behalf of the contractor, I certify that this Report is complete and correct and all equipment and material used and work performed during this Reporting period are in compliance with the contract plans and specifications, to the best of my knowledge, except as noted above.

John Davis	06/20/2025	*DN*	06/20/2025
QC Signature	**Date**	**Superintendent Initials**	**Date**

APPENDIX 10

CONTRACTORS QUALITY CONTROL REPORT (QCR)

Daily Log of Construction

Project	Fort Valor Command and Control Center	Report Number	QCR-2025-001
General Contractor	Horizon Builders, Inc.	Date	06/20/2025
Contract Number	FVC-2025-001		

Weather
▪ Weather Caused Delay: No
▪ Temperature: Min 60°F / Max 75°F
▪ Rainfall/Precipitation: 0.00 in (None)
▪ Wind Speed: 5 MPH

QC Narrative:
Today's inspection confirmed that all temporary facilities and security measures have been established in accordance with project specifications and federal security standards. The exterior secure cladding installation on the Secure Operations Center was observed for compliance. A minor deficiency in the placement of temporary security signage at the main entrance was identified and corrected within the day. All work is proceeding as scheduled with no critical deviations noted.

Prep/Initial Dates

Mobilization & Site Setup initiated: 06/15/2025

Activity Start/Finish

Mobilization & Site Setup: Start 06/15/2025, Finish (anticipated) 06/20/2025

QC Requirements

- Daily site inspections
- Verification of temporary facility setup and secure utility coordination
- Compliance with safety, security protocols, and project specifications

Issued QA/QC Deficiencies

Temporary security signage not correctly located at the main entrance

Corrected QA/QC Deficiency

Signage repositioned to specified location on 06/17/2025

General Contractors on Site

Horizon Builders, Inc. personnel (Installation crew, supervisors, and QC staff)

Labor Hours			
Employer	**Labor Classification**	**Number of Employees**	**Hours Worked**
Horizon Builders, Inc.	General labor	5	40
Horizon Builders, Inc.	Skilled trades	3	24
Horizon Builders, Inc.	Site supervisor	1	8
		Total: 9	**Total:** 72

Total Hrs Worked to Date: 72 hrs   **Hrs Worked MTD**: 72 hrs

Equipment Hours

Serial Number	Description	Idle Hours	Operating Hours
EQ-001	Mobile crane		
EQ-002	Forklift	2	6
		0	8
Total		**2**	**14**

Total Operating

Hrs to Date   14 hrs   **Idle Hrs MTD**   2 hrs   **Opr Hrs MTD**   14 hrs

Mishap Reporting

No mishaps reported.

General Contractor Certification

On behalf of the contractor, I certify that this report is complete and correct, and that all equipment and material used and work performed during this Reporting period are in compliance with the contract plans and specifica-tions, to the best of my knowledge, except as noted above.

John Davis	06/2/2025	DN	06/20/2025
QC Representative Signature	**Date**	**Superintendent Initials**	**Date**

SUPERVISOR'S REPORT OF INJURY

Name of Injured:	Timothy Gaines
Injury Date:	06/12/2025
Time of Injury:	10:15 am
Number of Lost Days:	3

Where and how did the accident occur?

On the morning of 06/12/2025, Timothy Gaines was unloading materials at the material staging area for the Fort Valor Command and Control Center project. While securing a heavy pallet, he slipped on a water-accumulated surface near the staging area. The spill had not been promptly addressed, resulting in a fall that caused a sprained ankle.

Unsafe act or condition:

- **Unsafe Act:** The injured worker was not wearing non-slip footwear despite known wet conditions.
- **Unsafe Condition:** The staging area had an accumulation of water that created a hazardous surface.

Measures taken to prevent a similar type of accident:

- Immediate cleanup of the water and implementation of enhanced drainage in the staging area.
- Enforcement of a mandatory non-slip footwear policy for all workers in wet areas.
- Additional training on safe material handling and hazard recognition was provided to all site personnel.
- Installation of warning signage in areas prone to water accumulation to alert workers of potential hazards.

Note: The person filling out this form must adhere to OSHA standards and/or complete the OSHA 300 Form, as applicable, to ensure compliance with all regulatory requirements.

Jessica Miller	*Jessica Miller*	06/12/2025
Author's Name	**Author's Signature**	**Date**

APPENDIX 11
US ARMY CORPS OF ENGINEERS

Deficiency Items Issued – by All

FVC-2025-001

Fort Valor Command and Control Center

06/20/2025

Fort Valor Military Base, VA

Item Number	Description	Location	Status	Date Issued	Age (days)	Staff
D-001	Temporary signage not installed per specs	Main Entrance	Open	06/16/2025	4	John Davis
D-002	Delay in material staging area setup	North Staging Area	In progress	06/17/2025	3	David Nguyen
D-003	Incomplete installation of safety barriers	South Perimeter	Resolved	06/15/2025	5	Maria Rodriguez

WARRANTY WORK ORDER

Project Superintendent:	David Nguyen	Site Administrator:	John Davis

Project:	Ft. Valor Command & Control Ctr.	Date W/O Requested:	01/15/2028
Requested By:	Jessica Miller		
Problem:	During a routine inspection of the Secure Operations Center's north corridor, several automatic sprinkler heads failed to activate during testing.		
General Contractor:	Horizon Builders, Inc.		

Date & Time On-Site:	01/17/2028, 09:00 AM	Client Notified:	☒ Yes ☐ No
Who Was Present:	John Davis (Sr. Project Coordinator), Mike Thompson (Field Technician), Maria Rodriguez (SSHO Manager)	**Name of Individual Notified:**	Jane Smith, Project Liaison

Work Accomplished:

- Performed full evaluation of the automatic sprinkler control valves and head assemblies in the north corridor.
- Recalibrated three control valves to factory settings.
- Removed and replaced four defective sprinkler heads with new NFPA-compliant units.
- Conducted system-wide pressure and activation testing; all heads now operate within specified parameters.

Date/Time Completed:	01/18/2028, 03:00 PM	Contractor Signature:	*John Davis*

Notes to Construction/Maintenance Personnel:
<ul><li>File completion notice with QA Department.</li><li>Update as-built drawings and O&M manuals to reflect valve and head replacements.</li><li>Schedule follow-up inspection in 30 days to verify continued compliance with NFPA and military fire safety standards.</li></ul>

APPENDIX 14
PRECONSTRUCTION CHECKLIST

Project #: FVC-2025-001

PROJECT FOLDER
- Drawings A/E (conceptual) 90%
    - Date Received: 05/05/2025
- Drawings A/E (final) 100%
    - Date Received: 05/15/2025
- C/D: ☒ Yes   ☐ No
- Matrix: ☒ Yes   ☐ No

PROJECT PROFILE
Scope of Work
Develop a state-of-the-art secure Command and Control Center on Fort Valor Military Base, VA. This project comprises two primary facilities:
- Secure Operations Center
- Administrative & Support Facility

The facility will integrate advanced security systems, secure communications, and robust support infrastructure to meet stringent federal and military standards. Designed to support 2,500 personnel at peak operational capacity, the completed facility is valued at $180 million. The project emphasizes the use of high-performance materials—including heavy steel, bullet-resistant glazing, and fire-resistive components—combined with state-of-the-art finishes and secure site logistics.

Utility Connections
- Electrical Connection: Connection to the base power grid via a designated secure service point on the site perimeter.
    - Location/Height: Approximately 15 ft above grade on the north side of the facility.

- Water Connection: Tied into the Fort Valor base water system via a secure connection point.
  - Location/Height: Connection located at the east side of the facility at grade level.

- Communications Connection: Provided by the base's secure high-speed data network.
  - Location/Height: Housed in a secure, weather-protected enclosure near the south boundary.

- Fire System Connections: Designed to meet federal fire protection standards for secure facilities with a wet/pre-action system per design.
  - Location/Height: Connection points are located at each facility's mechanical room level as specified in the detailed drawings.

- Drainage Connection: Tied into the Fort Valor base drainage system with inlets installed along the secure perimeter.
  - Location/Height: Located along the west perimeter (exact elevations as per design).

Site Conditions

1. Discrepancies: Minor variations in site topography have been observed compared to the initial survey, particularly in areas designated for secure access. Additional grading work may be required in these zones.
2. Differences between drawings and actual site conditions: Slight deviations in underground utility routing have been noted; coordination with the base utility services is underway to confirm final tie-in details.

Submittals

- % Submitted: 90%
  - Date: 05/20/2025
- % Comments: 10% pending review
  - Date Received: 05/21/2025
  - Date Responded: 05/25/2025

LONG-LEAD ITEMS

- Custom security glazing for the Secure Operations Center
- Specialized pre-cast concrete elements for the Administrative & Support Facility
- Long-lead secure HVAC and power distribution equipment

CONSTRUCTION ADMINISTRATION

1. M&H Representative conducting site visit: Mark Thompson (Materials & Handling Coordinator, Horizon Builders, Inc.)
2. Any job specific forms, instructions, and required procedures? Yes; all project-specific safety protocols, quality guidelines, and procedural instructions have been issued per the contract documentation.
3. Are drawings on site? ☒ Yes ☐ No
4. If no, specify which one and plans to get to site. N/A
5. Point of contact for clearance/access? Title of Representative? John Davis, Senior Project Coordinator, Horizon Builders, Inc.
6. What security clearances, if any, are required for the construction site? All personnel must have appropriate clearance as issued by the Fort Valor Military Base security office. Standard access badges and security training documentation are required.
7. Estimated delays entering and departing jobsite? Minimal delays expected due to controlled base access procedures.
8. Travel time to and from jobsite? Approximately 15 minutes from the base's central operations office.
9. General Contractor's Job Site Address: 100 Valor Way, Fort Valor Military Base, VA 22301
10. General Contractor Staging Materials: ☒ Yes ☒ No
11. If no specify General Contractor provided labor for material movement: N/A
12. Is governing agency providing secure adjacent area provided: Yes; Fort Valor Military Base provides a secure staging area adjacent to the construction site.
13. Walk material staging route/location: Materials will be staged at the rear of the facility at 100 Valor Way, with a designated secure route marked for deliveries.

14. General Contractor's marshaling point and secure shipment address: Marshaling point is at the main entrance of 100 Valor Way; all shipments are to be directed to this secure address.
15. Customers date for identification of installers coming on site: June 15, 2025
16. Customers agreed upon weekly conference: Every Monday at 8:30 AM via video conference

Point of Contact

- Architect's Representative
  Name/Job Title/Phone Number/Email

- Customer Representative
  Name/Job Title/Phone Number/Email

- Superintendent
  Name/Job Title/Phone Number/Email

- Client QC Inspector
  Name/Job Title/Phone Number/Email

- M&H QCM
  Name/Job Title/Phone Number/Email

- Logistics POC
  Name/Job Title/Phone Number/Email

Hotels/Apartments in the Area
[Local Hotels/Apartments/Addresses/Phone Numbers]

Transportation
Taxi and shuttle services available; estimated cost per night is approximately $250.

Hotel Services Cost
- Laundry: Approximately $25 per load
- Taxi: Approximately $30 per ride
- Room Service: Average $50 per order

- Phone Service: Local calling at approximately $0.10 per minute
- Transportation (to/from jobsite/airport): Approx. $40 per trip (airport shuttle service)

Medical Services (Location and Services Offered):
[Nearest facility/Address/Services Offered]

Special Safety/Security Concerns
Enhanced security protocols due to military base requirements; strict adherence to federal safety and security standards during deliveries and staging.

Charges for Expenses (Hotel/Transportation/Visa/Medical/Dental):
Expenses will be charged per actuals as outlined in the contract guidelines.

What Environmental Protection is Required?
Dust control measures, water runoff controls, and protection of adjacent green spaces per local environmental regulations.

CONSTRUCTION PROCESSES
- Planned Start/Completion Dates
  Start: July 1, 2025
  Completion: December 31, 2027

- Material Shipping Dates
  Critical long-lead items (e.g., custom security glazing, specialized HVAC components) are scheduled for shipment in early June 2025; remaining materials will ship beginning in early July 2025.

- Known Interruptions
  Potential delays due to weather conditions and base security protocols; contingency plans are in place.

- Outstanding Questions Regarding Plans/Specifications or Execution of Work
  Clarification required on final utility tie-in details and confirmation of secure access routes.

- Will Utilities be Installed Upon Our Completion of Work?
  Utilities will be installed by the base utility department post-construction; however, the contractor is responsible for coordinating interim utility connections as needed.

- Have all permits been acquired?
  - Building Permits: Yes
  - Utility Permits: Yes
  - Hauling Permit: Yes
  - Burn Permits: No (not applicable)
  - Any Other Permits: All required permits have been applied for and are under review.

- What Operation, Maintenance, and Certifications Must be Provided to the Customer?
  Operation and maintenance manuals, as-built drawings, and certifications for HVAC, electrical, and security systems are required upon project closeout.

- Problems with Submittal/Report Requirements?
  None identified to date.

- Special Quality Control Concerns
  Strict adherence to secure facility design specifications and regular quality control inspections for all materials and workmanship.

- General Contractor and Vendor Supplied Items
  All staging materials, temporary facilities, and essential construction equipment are supplied by Horizon Builders, Inc.

- Is Non-Organic Technical Assistance Required, and if so, what Arrangements Have Been Made?
  No additional technical assistance is required beyond the established project team.

- Will This Project be Affected by Priorities of Other Projects/Functions?
  No; the project schedule has been coordinated to avoid conflicts with other functions.

- What is the Schedule for Required Utility Outages?
  Utility outages will be coordinated with the base utility department; outages are expected to be minimal and scheduled during off-peak hours.

- What Arrangements Have Been Made for the Connection of New Utilities to Existing Service?
  The contractor will work closely with base utility providers to ensure seamless integration. Detailed connection plans are included in the project documentation.

- What Provisions Have Been Made for Temporary Utilities Service?
  Temporary water, power, and communications services have been arranged for the entire construction period.

- Project Manager/Superintendent Meeting Notes
  Weekly meeting scheduled; first meeting planned for June 15, 2025.

- Project Manager/Project Director Meeting Notes
  Monthly review meetings scheduled; first review set for July 15, 2025.

- Other Comments/Remarks
  All preconstruction items are on schedule. Any discrepancies or clarifications will be addressed during the weekly project meetings.

Jessica Miller	Project Manager	05/25/2025
Signature	Title	Date

PREPARATORY CONTROL WORKSHEET

Fort Valor Command and Control Center   06/15/2025

FVC-2025-001

Fort Valor Military Base, VA

Definable Feature of Work: [Definable Feature of Work]
A. Activities Included Under [Definable Feature of Work]
General Contractor – [RESP CODE]

A-001	Site Clearance & Demobilization of Debris
A-002	Installation of Temporary Facilities (Secure Site Office, Storage)
A-003	Establishment of Perimeter Fencing, Secure Access Controls, and Safety Signage

B. Quality Control Requirements
▪ All mobilization activities must comply with project specifications and federal security standards.
▪ Daily QC inspections will be conducted for staging, material handling, and temporary facilities setup.
▪ QC reports will be submitted to the Project Manager and reviewed for corrective action if required.

C. QA/QC Deficiency Items
▪ No significant deficiencies noted on initial review.
▪ Minor adjustments to temporary signage placement to be recorded in the daily report.

D. Labor Rates

Labor Classifications	Basic Rate	Fringe Benefits	Plus %	Total Wage/ HR
General laborer	$25.00	$5.00	10%	$33.00
Skilled tradesman	$40.00	$8.00	10%	$52.80
Site supervisor	$50.00	$10.00	10%	$66.00
Equipment operator	$35.00	$7.00	10%	$46.20

E. Review Contract Drawings and Specification

	Discussed		
	Yes	No	N/A
1. General site layout	X		
2. Utility connection details	X		
3. Temporary facilities and staging specifications	X		
4. Safety and environmental requirements for a secure facility	X		

F. Repetitive Deficiencies Found on Previous Projects

	Discussed		
	Yes	No	N/A
1. Improper signage installation			X
2. Incomplete staging area setup			X
3. Inconsistent material handling procedures		X	
4. Delayed permit and security clearance approvals			X

G. Control Checks

	In Compliance		
	Yes	No	N/A
1. Site clearance completed	X		
2. Temporary facilities in place	X		
3. Secure Utility coordinator confirmed	X		
4. Safety and security barriers installed	X		

H. Job Site Safety

	In Compliance		
	Yes	No	N/A
1. PPE availability	X		
2. Emergency exits and secure routes marked	X		
3. Hazardous areas identified and secured	X		
4. Daily safety and security briefings conducted	X		

I.	Quality Assurance Evaluation Notes			
		Discussed		
		Yes	No	N/A
1.	All QC inspections scheduled and documented	X		
2.	Material verification in progress	X		
3.	No major deficiencies observed	X		
4.	Minor rework required on temporary signage	X		

PROJECT FLOW CHART

	PRECONSTRUCTION PHASE	Responsibility
1	Prime Contract Awarded (LOI)	Business Development
2	Issued for Construction Drawings Completed	DM/PM
3	Estimating Binder (All Scopes Included & Take Offs) Completed	Chief Estimator
4	Construction Kick-Off Meeting	ALL DB Staff
5	Paperwork Forwarded to Client (SF 1413, COI, etc.)	Admin
6	Review Permit, Licensing, and Security Clearances for All Phases; Approve Drawings from Governing Agency	PM/APM
7	Obtain Bond & Forward to Client	PM/Accounting
8	Prioritize, Clarify Scopes & Buy-Out Project According to Construction Process & Long Lead Times	PM/Estimator
9	Obtain Proposals from Subcontractors (Revised Proposal to Reflect Buy-out)	PM/Estimator
CONSTRUCTION (3 Sections Mobilization, Construction, Close-Out)		
1-Mobilization		
10	Identify and Assign Site Staff to Project	Director
11	Create PMs Project Package & Financial Binder	APM
12	Dispatch Superintendent & QC Manager Package to Site	APM
13	Set-up Secure Project Trailer	SUP
14	Set-up Temporary Secure Bathroom Facilities	SUP
15	Erect Secure Project Signage	SUP
16	Set-up Dumpster	SUP
17	Coordinate Temporary Water Supply	SUP
18	Coordinate Temporary Gas Supply	SUP

19	Coordinate Temporary Power Supply	SUP
20	Coordinate Temporary Telephone/ Communication Services	SUP
21	Establish Project Construction Schedule (Level III)	PM
22	Create Daily Project Reports, Milestone Tracking, and Level II Sitrep	PM
23	Verify Scope, $ Value, Submittal Completion, Close-Out Documents Specified in Subcontracts	PM
24	Revise Subcontract to Encompass All Changes	PM
25	Award Contracts to Subcontractors	PM
26	Obtain Fully Executed Contracts	Contract Admin
27	Obtain Certificate of Insurance and W9 from Subcontractors	PM
28	Request Submittals from Subcontractors	QCM
29	Obtain Approval for All Submittals	QCM/DM/PM
30	Conduct Pre-Construction Conference with Customer, DM, DB Director, PM, Superintendent, and QCM	ALL
31	Conduct Pre-Construction Conference with Subcontractors, Superintendent, PM, and QCM	ALL
32	Identify Long Lead Items and Tracking Process	PM/SUP/QCM
CONSTRUCTION (3 Sections Mobilization, Construction, Close-Out)		
2-Construction		
33	Verify Existing Site Conditions	PM/SUP
34	Commence Construction of Project	PM/SUP
35	Monitor Long Lead Items (Periodically)	PM / SUP / QCM
36	Update As-Built Drawings (Periodically)	SUP/QCM
37	Coordinate Project Inspections & Testing	SUP/QCM
38	Complete Daily Project Reports, Safety Reports, Photo Sitrep, and Level II Updates	PM/SUP
39	Hold Weekly Meetings with Customer, Superintendent, A&E, and PM	SUP/QCM
40	Provide Cost Forecasts (ETC)	PM
41	Complete Monthly Pay Applications	PM/SUP

42	Submit All Change Order Requests (CORs) to Owner	PM
43	Obtain Owner's Approval for All CORs	PM/Owner
44	Achieve Project BOD (Beneficiary Occupancy Date)/Project SC (Substantial Completion)	PM/SUP
45	Superintendent & QCM creates M&H Punch List and Correct Items with Subcontractors	SUP/QCM
46	PM Completes M&H Pre-Punch List and Superintendent Corrects Items with Subcontractors	PM/QCM
47	A/E Completes Punch List and Superintendent Corrects Items with Subcontractors	SUP/A-E/QCM
48	Director Completes M&H Pre-Punch List and Superintendent Corrects Items with Subcontractors	DIRECTOR/QCM
49	Customer Completes Punch List and Superintendent Corrects Items with Subcontractors	SUP/QCM/ Customer
50	Complete All Punch List Items	SUP/QCM
51	Coordinate Move-out Efforts	SUP
CONSTRUCTION (3 Sections Mobilization, Construction, Close-Out)		
3-Close Out		
52	Project BOD (Building Occupancy Date)	SUP
53	Obtain Final Contract Value Change Orders from Subcontractors (Noting Contract Value Agreement)	PM
54	Obtain Superintendent/QCM Files	APM
55	Send Completion Letter to the Customer, Signifying Commencement of Warranty	Business Development
56	Obtain Letter of Recommendation and CASS Rating from Customer	Business Development
57	Complete Financial Close-Out (Final Application to Accounting then to Owner)	PM
58	Complete As-Built Drawings	APM

59	Compile & Submit Close-Out Documents to Customer	APM
60	Compile & Submit All Close-Out Documents to User	APM
61	Complete M&H Project Debrief (In-House)	ALL
62	Complete Subcontractor Debriefing	ALL
63	Complete Customer/Architect/Engineering/PM Debriefing	ALL

REDICHECK PLAN & SPECIFICATION REVIEW

Preliminary Review
☐ Quickly make an overview of all sheets, spending no more than one minute on each sheet, just to become familiar with the project you are assigned.

Specification Check (Technical Binder)
☐ Check specs for items sold. Do they coordinate with drawings?

☐ Check specs for phasing of work. Does our work fit into schedule appropriately?

☐ Ensure all finishes are specified in specs.

☐ Check major items to verify coordination with drawings. Pay particular attention to rating (i.e. Hertz, Voltage, CFM, etc.).

☐ Verify items specified and verify they are on drawings.

☐ Verify that cross references in specification do verify.

Plan Check
☐ Verify all dimensions.

☐ Verify plans match BOM on drawings.

☐ Verify most up to date approval on drawings.

☐ Verify electrical and mechanical match architectural.

☐ Verify all major electrical/mechanical items match BOM.

☐ Verify all notes.

☐ Verify dimensions are realistic in perspective to parent rooms.

☐ Verify subcontract and general contractor requirements.

ESTIMATING WORKSHEET

Estimating Worksheet						
Project #:	FVC-2025-001	**Customer:**	US Army, Fort Valor	**Date:**	06/25/2025	
Activity #:	001	**Prepared by:**	Jessica Miller [PM]			
Drawing #:	A-15	**BM#**	BM-01			
Project Location:	Fort Valor, VA	**Checked by:**	David Nguyen [SUP]			

Description	BM Item #	Issue Unit	QTY	Remarks
Site Clearance & Security Perimeter Setup	BM-001	LS	1	Includes removal of debris and installation of secure fencing and barriers along the site perimeter.
Temporary Facilities Setup (Site Office and Storage)	BM-002	LS	1	Setup of a secure temporary site office and storage container with necessary utilities.
Access Control & Surveillance System Installation	BM-003	LS	1	Installation of access control systems, surveillance cameras, and associated control panels along the perimeter.
Utility Coordination & Temporary Power Setup	BM-004	LS	1	Coordination with base utility department for interim power supply and utility connections.

Security Glazing for Operations Center	BM-005	SF	5,000 sq ft	Installation of high-security, bullet-resistant glazing as specified in drawing A-15 for the Secure Operations Center.

SAMPLE PRE-BID MEETING MINUTES

Date:	05/22/2025
From:	Jessica Miller
RE:	Fort Valor Command and Control Center, Project No. FVC-2025-001
Subject:	Pre-Bid Meeting Minutes

Project Team Introduction	
Owner	Jane Smith, US Army, Fort Valor
Project Manager	Jessica Miller
Construction Manager	David Nguyen

Project Summary

The Fort Valor Command and Control Center is a green-field project located on Fort Valor Military Base, VA. The development comprises two primary facilities—a Secure Operations Center and an Administrative & Support Facility—integrated with advanced security systems, secure communications, and robust support infrastructure. With a contract value of $180 million and designed to support 2,500 personnel at peak operational capacity, the project emphasizes state-of-the-art secure design, sustainable construction practices, and efficient site logistics. Key design features include comprehensive access control and surveillance systems integrated throughout the facility.

Anticipated Addenda Items

An addendum will be issued at least seven days prior to the bid date. Expected topics include:

- Revised utility connection details (electrical, water, communications)
- Updated site grading and staging area requirements
- Clarifications regarding temporary facilities and material handling protocols

Amendment #1	
Project:	Fort Valor Command and Control Center
Project No.:	FVC-2025-001
Date:	05/28/2025

Acknowledge Receipt of Amendment	
Signed:	John Davis
Title:	Senior Project Manager
Company:	Horizon Builders, Inc.
Date:	05/28/2025

Amendment

Amendment #1 clarifies the following:

- **Utility Connections:** Updated connection points for electrical, water, and communications have been provided. Revised details address minor discrepancies identified between initial base surveys and submitted drawings.
- **Site Conditions:** An updated grading plan reflecting slight variations in topography within the secure access areas has been included.

- **Staging Areas:** Revised requirements for material staging and temporary site facilities have been outlined to ensure smooth operations during mobilization while adhering to base security protocols.

General Contractors are advised to review the attached addendum documents carefully and adjust their bids accordingly.

FIELD ADJUSTMENT REQUEST (FAR)

FAR Number:	FAR-001	Page:	1 of 1
Project Number:	FVC-2025-001	Limit Number:	LMT-01
Submitted By:	Jessica Miller	Date:	06/20/2025

Description of and reason for request (include drawing and sheet numbers and attached drawings as necessary for description):

During installation of the secure cladding panels on the north elevation of the Secure Operations Center, discrepancies were noted between the field conditions and the approved drawings (Drawing A-15, Sheet 3). Specifically, the measured mounting bracket spacing deviates from the design specifications, likely due to unforeseen variations in the secure construction environment at Fort Valor Military Base. A field adjustment is required to realign the cladding installation to conform with the approved design. Revised drawings and updated calculations are attached for review.

Route to	Approved: (initial)	Denied: (initial)	Date:	Estimated additional cost: $10,000
Installer	JD		06/21/2025	
PM	JM		06/21/2025	Estimated additional days: 10 days
Eng.	DL		06/21/2025	
GC	DN		06/21/2025	
OBO	JS		06/21/2025	As built: JM (initial)
PM Signature	*Jessica Miller*			Date: 06/22/2025

Notes

1. Route original and 3 copies to Owner's Representative
2. Owner's Representative returns original and 2 copies

FIELD ADJUSTMENT REQUEST (FAR) SUBMITTAL LOG

FAR Number	Description	Spec Section	Drawing Number	Date to GC	Date Returned	Approved/ Denied
FAR-001	Exterior Secure Cladding Panel Adjustment	07 24 00	A-15	06/20/2025	06/21/2025	Approved
FAR-002	Revised Secure Utility Connection Details	22 10 0	E-12	06/22/2025	06/23/2025	Approved
FAR-003	Secure Access Area Grading Adjustment Request	03 30 00	S-08	06/25/2025	06/26/2025	Denied

CHANGE NOTICE REQUEST

		Change Order Submittal Log				
CO#	Description	Spec Section	Drawing Number	Date to/ from GC	Date Returned	Approved/ Denied
CO-001	Revised Secure Cladding Panel Installation Detail	07 24 00	A-15	06/22/2025	06/23/2025	Approved
CO-002	Additional Site Grading Work	03 30 00	S-08	06/23/2025	06/24/2025	Approved
CO-003	HVAC Duct Rerouting Modification	23 05 00	M-03	06/24/2025	06/25/2025	Approved

www.ingramcontent.com/pod-product-compliance
Lightning Source LLC
Chambersburg PA
CBHW031806190326
41518CB00006B/221